LUNA
The Astrological Moon

Paul F. Newman

The Wessex Astrologer

Published in 2011 by
The Wessex Astrologer Ltd
4A Woodside Road
Bournemouth
BH5 2AZ
England

www.wessexastrologer.com

Copyright Paul F. Newman 2011

Paul F. Newman asserts the right to be recognised as the author
of this work

ISBN 9781902405575

A catalogue record of this book is available at The British Library

Cover design by Tania at Creative Byte, Poole, Dorset

Cover picture: *Oberon, Titania and Puck with Fairies Dancing*,
by William Blake, 1786. (Detail).

All rights reserved. No part of this work may be used or reproduced in any manner without written permission. A reviewer may quote brief passages.

'Of cloudless climes and starry skies;
And all that's best of dark and bright
Meet in her aspect and her eyes:
Thus mellow'd to that tender light
Which heaven to gaudy day denies'
    [Byron. *She Walks in Beauty*]

'My mistress' eyes are nothing like the sun...'
    [Shakespeare. *Sonnet 141*]

To the memory of Coco

My little moon

CONTENTS

A Lunar Prowl	1
The Pale Fire	3
Introduction	3
The Marriage of Sun and Moon	4
The Moon is not Venus	7
Fertility and Pregnancy	8
Psychic Tendencies	9
Water and the Moon	10
Winds of Change (summer and winter moons)	11
The Goddess with Three Faces	19
The Number Three	21
The New Moon	31
Some New Moon charts:	33
(The Paparazzi, First 45rpm record, Darwin's *Origin of Species*)	
The Crescent Moon symbol	38
Lunisequa – the Lunar Star	41
The Full Moon	43
Special Full Moons	44
Bewildering Lunations	46
The Moon's a Crowd	48
Some Full Moon charts:	49
(Discovery of DNA, Eurovision Song Contest,	
Benny Hill, Lara Croft)	
The Old Moon	57
The Cinnamon Tree in the Moon	58
Some Old Moon charts:	59
(George Orwell, *Punch*, Viagra)	
The Moon was Pluto: The Dark Side of the Moon	64
The Witch	65
The Computational Moon	75
Moon Maths	77
The Moon's cycles	78
The Lunar Zodiac: The Mansions of the Moon	82
Incongruous Bookends: Magic Squares of Moon and Saturn	87

The Tarot Card of the Moon	91
The Progressed Moon	96
Moon hours	98
Zodiac degrees with moon relevance	100
The Moon as final dispositor: *Casablanca*	100
Eclipses:	103
The Solar Eclipse. 'God is supplanted'	103
Pink Floyd and the 1999 Eclipse	106
The Lunar Eclipse. 'A goddess in purdah'	109
Bad Moon Rising	111
The Poetic Moon	115
Moonscapes	117
The Moon in Art	117
The Moon as Muse	118
Mermaids	122
Rising Moons – The Ship's Figurehead	124
The Moon and Fairy Queens	127
The Owl and the Pussycat	129
Vintage Moon Fiction	133
The Black Moon	153
The Black Moon Lilith	155
Priapus	159
Lilith, the history and legends	159
The Dark Goddess of Nature	164
Duende	164
Cats	165
Black Moon Nakshatras	167
Film Noir and Fatal Women	168
Sunset Boulevard	169
Conclusion	173
Why are they ruled by the Moon?	173
Appendix: How to draw a Crescent Moon	185
Index	187

A LUNAR PROWL

'… all ye that walk in Willow-wood,
that walk with hollow faces burning white…'
  [Dante Gabriel Rossetti, *Willow-wood*]

The Pale Fire

> 'The sun's a thief, and with his great attraction
> Robs the vast sea
> The moon's an arrant thief
> And her pale fire she snatches from the sun'
> [Shakespeare, *Timon of Athens*]

Introduction

In astrology the Moon plays equal partner to the Sun:

> She is the Lady of the Night to the Lord of the Day.
> They are Mother and Father of a shared domain.
> Yin and Yang.

This observation is only clear to those who were born and live on earth. The sun and the moon are the largest lights in our sky and appear to us as equal size. The golden god rules the day, the silver goddess rules the night. And that balance is so fine that life on earth would have evolved very differently without it.

On those occasions when the sun and moon totally embrace each other as at a solar eclipse, the balance becomes delicate and the earth trembles. Nature is momentarily confused and the world goes dark.

Yet in many branches of astrology, including horary, medical and Vedic, the Moon is a more important planet than the Sun because it has a swifter movement. The moon is periodically a greater umbrella than the sun, so to speak, because it can be overhead at 28+ degrees of terrestrial latitude either side of the equator, whereas the sun only reaches its overhead limit at 23 degrees 26 minutes north and south, the Tropics of Cancer and Capricorn. The moon has a broader spread across the earth if viewed in this way. However it is typical that while the sun will make its overhead journey regularly each year, reaching its midsummer heights and midwinter lows (the solstices or standstills), the moon will only outdo its partner sometimes and not others. The moon's major and minor standstills are separated by intervals of 9-10 years.

By its nature the moon is not amenable to clear-cut intellectual analysis. It does not often reveal itself best in that manner. In this book I have avoided making lists of astronomical facts about the moon or lists of moon goddesses from different cultures that can be readily found elsewhere. This is not an encyclopedia of the moon. But it will be necessary, if not essential, for us to wander into a little poetry from time to time, to graze in the imaginative fields of literature and permit ourselves to wallow in nostalgia – although my own Uranus opposite Moon will probably not allow too much of this. I hope you will find some New insights in the following pages, along with some Old, Full and Black moon insights.

I have also had to decide whether I should write the word Moon with a capital letter or with a small letter. You will already find it wavering in the paragraphs above. In the end I have attempted to use the capital letter M when referring to the Moon on a particular astrological chart and a small letter m when speaking of the moon in general. Unfortunately many times you could argue it either way.

So let us honour the M(m)oon in all her habitual inconstancy.

The Marriage of Sun and Moon

The moon is the beautiful wise one, protecting and nourishing mother to all. The moon is the powerful memory, real or imagined, of the mother's breast at which we find shelter and sanctuary from grief. The secreting 'milk' of human kindness. Our Moon sign contains that security. But she can also be a cold-hearted bitch and a sob-story slut.

The cold moist moon complements the searing clear-sighted sun by offering a different view, a night-time view where our needs, urges, appetites, and senses other than sight take precedence. If the moon is termed irrational, it's only because we are judging it from the rational yardstick of the sun.

We all perceive the world through both these eyes of sun and moon and blend them automatically. It remains one of the oldest and most profound of wisdoms that the left eye of a man is ruled by the moon, and the right eye of a woman is ruled by the moon. (And the left eye of a woman is ruled by the sun and the right eye of a man is ruled by the sun).

The moon fits comfortably into what has become known as right-brain activity; although the whole concept of 'left and right' is confusing to the moon which naturally reflects everything anyway. But the similarities are there. The moon is creative and imaginative, it engages the hunger and the feelings; it employs music and symbols in a world not bound to the usual structures of time. That we can often literally observe the moon in the daytime but never the sun at night is symbolic of Luna's looser overlapping fluidity.

In the popular almanacs of previous centuries it was the changing position of the moon rather than the sun that gave the basis for prediction. The varying phases of the moon and its effect on sea tides and agriculture were observable to everyone, regardless of learning, race or culture.

Astrology infers that women are ruled by the moon, and it must therefore be more personal in a woman's map. A woman's moon refers not only to her mother but to herself as a possible mother. A man's moon refers primarily to his mother, but as he cannot biologically be a mother it refers to all women in his life and his relationship to everything lunar. A man naturally sets about his business like the sun; he can never be the moon in the sense that women can, but he can seek to find it in various ways. We should remember too that whether you are man or woman your Moon is ultimately a part of you. It's not your mother, she's got her own Moon; it's not your wife, she's got her own Moon. She (the Moon on your chart) contains important facets of *your* personality. She's not an alien thing that belongs to somebody else. She's yours.

To say that women are ruled by the moon suggests that men are ruled by the sun, but is either assumption wholly true? Luna is a feminine name, like Diana, Maria, and countless names ending in 'a'. But determined by the different views from our eyes the two sexes may be genetically primed to emit and receive through the windows of their soul the balance of solar and lunar differently. Only when a man and a woman look into each other's eyes do their suns and moons match up.

'Our eye-beames twisted, and did thred…' [1]

Your reaction to the above paragraphs, and in fact to the whole of this book, will be governed by the sign and condition of your Moon.

The moon in astrology offers a feeling of security; often the security of a long ancestry known or unknown that stretches back in time. Almost everyone finds a security in the foods of the culture in which they were brought up; the moon rules nourishment and the sign of Cancer rules the stomach and the moon nourishes both our bodies and our feelings with its comfort food. Many find security in their faith or religion or have been raised in a certain faith through their family background. If this applies your Moon should confirm it and tell the story. Yet for men or women the personal moon is not just an eternal comforter, it also seeks our authority. It's given light by our leadership, by our sun-will. It needs input from us to

---

1. John Donne. 1572-1631. From *The Extasie*.

reflect back insights and intuitions, though these will come in her own time and may seem, when they come, unbidden. We mustn't be afraid of our Moon; it thrives on our direction and willpower. It responds to direction because it has incredible powers to earth and actualise. In other words your Moon needs you just as much as you need her. You must *know* her or she will use you, in the sense that you will be forever at the mercy of your instincts, acting blindly without fully realising what you are doing. Without putting it too strongly, you will be acting like a lunatic.

Alternatively, as the saying goes, we should be careful what we wish for. Or, we might add, beware of what we wish for with passion for then we really do set forces in motion. A demand or statement borne and delivered with intense emotion will draw the moon into it. When Scarlett O'Hara in *Gone With the Wind* raged defiantly to the universe "I'll never be hungry again!" we recognised a form of positive affirmation (albeit in negative construction). And so great was the heat and force of that demand that she never was hungry again. It is the basis of much modern New Age thought that to accomplish a positive result the request for a state or thing desired should be accompanied by a strong and focused emotion. The practice is as old as prayer. And whomever the deity addressed, if there is feeling behind it, the moon (your Moon) is brought into the equation. That way great and beneficial things are accomplished.

The dark side of this is the curse. (Not in this case the so-called 'curse of Eve', the menstrual period, but a verbal curse). A verbal curse is a negative demand delivered with high emotion, and that has the moon's power too. The results in this case are destructive, but the moon itself is not to blame. She is a planet like any other in astrology. The energy she symbolises and emits can swing either way.

Understanding your Moon is an imperative to wisdom and self-knowledge. Do you really understand what it means to have the Moon in the sign and in the house that it tenants on your chart – and the main aspects to it? In what way does this make you feel at home or secure? Or doesn't it?

We can also help our lunar understanding by deliberately striving to blend our personal Sun and Moon together. Focusing on the degree of their zodiac midpoint is one way of doing it – a solar way. Or we could do it in lunar fashion by integrating the pictures of our Sun and Moon signs and making a union of the two symbols. For example Leo and Virgo (lion and maiden) together make a Sphinx. Gemini and Pisces together might create a couple of Mermaids. Be imaginative and use some of the alternative symbols below if you find them appealing, or any others relevant to your sign.

> Aries is also a Lamb, a Red Ruby, a Raven, a Hawk.
> Taurus is also a Flower Garden, an Apple, a Cherry, a Copper Band.
> Gemini is also a Bird, a Hand, a multi-coloured Marble.
> Cancer is also a Ship, a Hare, an Egg, a Pearl.
> Leo is also a Crown, a Chrysanthemum, a Peacock, a Swan.
> Virgo is also an Angel, a Bee, an ear of Corn, a sprig of Lavender.
> Libra is also a Dove, an Emerald, a Peach.
> Scorpio is also Serpent, an Orchid, a Pomegranate, an Eagle.
> Sagittarius is also an Acorn, a Conquistador, a Purple Silk.
> Capricorn is also an Anchor, a Unicorn, an Ivy wreath.
> Aquarius is also an Airplane, a Glass Bead, a Lightning Flash.
> Pisces is also an Amethyst, a Pelican, a delicate Shoe.

The Moon is not Venus

Feminine attributes other than motherhood, the nurturing function and general receptivity, are ruled by Venus rather than the Moon.

Peaches and cream are not the Moon's cuisine; neither do the pink rose petals of love and flirtation flutter down from Luna. The Moon may join in Venus' amorous mix to stir deeper feelings like the need for security, or an unreasonable jealousy or the melancholy of the lost or unrequited love. From nostalgic sentimentality to utter despair the Moon is then in command. But the force of attraction between sexes belongs to Venus, as do all close unions and relationships; as does beauty, music, fashion, art… and erotica.

Mistress of the Night is an awkward term for the moon because of the double meaning in English of the word 'mistress'. Intended as the feminine equivalent to the sun as Master of the Day, Mistress of the Night simply means the ruler of the night. But in today's world we rarely use the word mistress to signify a woman in authority; it is more commonly understood to mean a woman who is in an illicit sexual relationship with a man. Thus we hark back to the old idea that the moon was dangerous to morally upright men because it sent uninvited female images or spirits to tempt him at night. But the moon's part in this was not as a generator of erotic fantasies (that would be Venus), it was simply a mirror to everything consciously suppressed from the waking mind. An echo perhaps of the myth of Endymion who was loved by the moon goddess but could only be visited by her in his sleep.

And the moon could be unreliably dangerous to both women and men as there are sides to her that are never constant. The moon can be fickle, though in another typically contradictory way it can harbour prejudice and rule those unchanging ingrained habit patterns that we follow blindly and routinely without ever questioning.

The Moon is like the sea. Predictable in its tides, unpredictable in its force.

Like the sea, this changeable tidal influence is most notably experienced from the land, or where the land meets the sea – the shoreline. The ebbing/flowing meeting point of worlds is similar to the Earth's relationship with its closest planetary influence, the shifting silvery moon. The moon is closely tied to the Earth – in all ways. It rules nature and the patterns of fertility; it is exalted in the earthiest of Earth signs, Taurus. Both the Moon and the Earth are seen as feminine and both are seen as mothers. The sun may rule the changing seasons and quicken the seeds of life, but it is the moon that produces the eggs and seeds and nurtures them in the protective darkness of womb and soil.

Fertility and Pregnancy

Menstruation and the moon are inextricable. The word 'menstruate' comes from the Greek *mene*, meaning moon. To the Ancient Greeks the menae were the goddesses of the lunar months, female children of Selene, the moon. (The word 'month' also comes from 'moon'). We might ask, if the moon were situated further away from the Earth and took longer in its orbit would a woman's monthly period then only happen every three months, or every six months, or longer?

The first of the following two rules on fertility has long survived through Old Wives' Tales, a natural dark moon function. So although no guarantees are attached and the usual disclaimers that this is not intended to overrule a qualified medical doctor's advice are hereby given, both methods are still worth passing on to anyone who might be trying to conceive a child.

The classic rule is that women are supposed to be at their most fertile each month when the moon's phase in the sky is identical to the moon's phase at their birth. In other words the sun and moon should be the same number of zodiac degrees apart as when a woman was born.

The second is more modern[2] and in line with regularising the menstrual period. Two weeks (14 days) after a woman's last period would be the

---

2. Credited to physicist Edmond Dewan in 1967.

equivalent of a personal 'full moon' of fertility and so to sleep with the light on all night at this time has been claimed to stimulate ovulation.

Astrologically a transit of Jupiter to the Moon on a woman's chart could be an indicator of pregnancy. And a birth chart with the natal Moon in the fifth house is usually an indication of fertility.

Combinations of the Moon and Venus, with one planet transiting or progressing to the other, can indicate conception or birth. Uranus with its quickening nature adds more to the picture – Uranus coming to the Moon/Venus midpoint or the Moon to the Venus/Uranus midpoint by transit or progression can be symbolic of the sudden revolutionary change of a new birth. Moon-Venus-Pluto combinations have similar motherhood properties, or the Moon with Mars and Jupiter.[3]

The constant factor throughout all this is the Moon.

Psychic tendencies

In several cultures the moon was believed to be the abode of souls or spirits; an ever changing waxing and waning clearing ground comprised of souls on their journey back to reincarnation or on to higher spheres. The solar system and the planets were thought the abode of angels, and the realm of the fixed stars the abode of archangels. Whatever your belief the sub-lunar realm, the area between the earth and the moon, understandably contains the residue of a million emotionally charged hopes and fears hanging around like forgotten wraiths close to the earth, and many unreliable messages and channellings can originate from here.

Everyone is 'psychic' to some extent and there are many special people who live continually open to a world of spirit. 'Psychic tendencies' in a horoscope reading are often given to Pisces or Neptune but true psychic ability must involve our Moon, the most receptive of planets.

Naturally amenable to the irrational the moon is mediumistic and will pick up surrounding atmospheres and invisible influences, behaving like a radio with a random rolling tuner. The Moon on the ascendant is particularly prone to this, as is the Moon closely aspected to an outer planet, where it becomes a receiver of outer influences. The outer planets and the moon share common bonds as many of the astrological characteristics and meanings of the outer planets belonged to the moon before their discovery. The sudden changeability of Uranus was the moon's fickleness, the Neptunian world

---

3. Reinhold Ebertin, *The Combination of Stellar Influences*, AFA, Tempe, 1992. Originally published in German in 1940.

of dreams and enchantment was that of Luna, and the dark underworlds of transformation and sunless taboos belonged to the dark moon before Pluto. [See more on this in 'The Moon Was Pluto']. Any conjunction or strong aspect to the moon from any planet tends to *earth* the moon and allow it to be actualised in a certain way. This is not to say that the moon only sends muddled unreliable messages, far from it; it can be a channel to your most reliable intuitions.

It may be misleading to equate the Moon on a chart with one's 'soul', but to speak of something having 'soul' captures that almost indefinable dimension of knowing and feeling that characterises the moon.

A distinction between Water and the Moon

The element of Water is that of emotion and of memory, and the moon that rules the tides of the sea is naturally associated with it. We cry salty watery tears through sadness, joy, self-pity, empathy, relief, thankfulness, nostalgia, sorrow, patriotism, pain… the 'depth' of our feelings is sometimes unfathomable.

There is now irrefutable evidence that there is water on the moon. NASA announced that 24 gallons of frozen water mixed with dust and debris was thrown up when the LCROSS satellite was crashed into the moon's surface in 2009. And later reports confirmed that craters near the moon's north pole contained millions of tons of ice.

Water, like each of the other elements, is neutral, it can refresh you or drown you and those who have a strong Water emphasis on their chart may cry more easily than those who do not. But that does not make them unhappier individuals than others, nor necessarily more compassionate.

The Moon however is not the total source of this. The Moon was, and still is in Vedic astrology, the mind. This is separate from the Mercury function of communicating thought and ideas. Mercury was the son of the Moon in Hindu myth, and we can see the commonly understood attributes of both the Moon and Mercury together in the Egyptian deity Thoth, who was both a night sun and a divine scribe. The intellect derives from a greater mind. The moon, which embraces the unconscious world, is basically an unconscious force (the moon is naturally dark) operating through our herd and survival instincts. The goddess Diana ruled over the animal world because animals live almost purely by instinct. Mercury separates us from this obedience and allows us to think and make choices. Mercury, in other words, makes us human. But the moon is the mind's direct awareness rather than its reasoned analysis. The moon is the mind's first overall impression; the gut feeling. The moon reveals its light to us in ever-changing phases, allowing access to unexpected depths and enlightenments of the personal and collective mind. In meditation the aim is to still the mind and reach a calm and tranquil centre, to pull back wandering thoughts and let them dissolve like clouds dispersing across the face of the moon. When our minds wander we may blame Mercury, and it may *be* Mercury if we are led into an active thought pattern. But if the mind conjures up random images or places and faces from long ago, again suspect the moon.

Habit, reaction and reflex are governed by the moon, and the Moon's sign shows what we instinctively need. When conscious defences are lowered, as perhaps when people are slightly drunk, most will show their Moon's sweet side – the soft spot. But others turn awkward and aggressive and have to apologise for their behaviour later, if they consciously recall it at all. Similarly if we are tired or low on energy we may be surprised at our snap reactions to otherwise trivial situations. This again is the moon temporarily taking over. It has many different faces.

Winds of Change:
Summer And Winter Moons

> Summer New Moons are Yang; Winter New Moons are Yin.
> Summer Full Moons are Yin; Winter Full Moons are Yang.

Boreas, the North Wind, was a fierce and impetuous blast of mighty force. He gave his name to the first six zodiac signs, the 'Boreal' or Northern signs, Aries through to Virgo. In contrast Auster, the Southern Wind, was a subtler, gentler, more complex soul and because of celestial geography held association with the Southern signs of Libra to Pisces.

The Northern signs represent the northern hemisphere of the Earth. The Sun moves into the northern hemisphere at the March equinox each year (the Sun's Ingress at zero degrees Aries) and stays in this northern declination for the next six signs and the next six months. It represents summer in the northern hemisphere of the globe. March equinox to September equinox. Aries through Virgo. This is the meaning of 'summer' in the following examination.

'Winter' for our purposes refers to the period from the September equinox to the March equinox. In late September the Sun enters the southern hemisphere and travels through the Southern signs of Libra through Pisces. (Naturally the actual seasons are reversed if you live in the Southern Hemisphere. Winter is summer and summer is winter. But the Sun's movement through the zodiac signs remains the same as does the symbolic qualities of North and South Wind.)

Throughout its course during both summer and winter the Sun will regularly conjunct and oppose the Moon, giving us new moons and full moons. There will be six new moons and six full moons in summer, possibly seven of one of them, and the same number in the winter.

If we think about zodiac signs it will be clear that during the summer 'North Wind' phase every new moon takes place in a Northern sign. That is, both the Sun and the Moon will be together in a Northern sign (Aries-Virgo) at new moon allowing an abundance of straightforward Boreal energy to accompany each Sun-Moon conjunction. Two weeks later when the Moon is at its height at full moon, the Moon will be in the domain of the South Wind. The Moon, now opposite the Sun, will be in a Southern sign. This full moon will therefore be less direct in its effect than the new moon preceding it. Full moons in the summer ride visually lower in the sky (they are higher in the sky in winter). So what begins on a new moon in the summer begins with a direct straightforward North Wind thrust, but when it reaches its lunar culmination at full moon fourteen days later the outcome is more of the nature of an inner illumination, more complex and diffused, more like Auster the South Wind. This inner understanding of northern motivations is born of an outer trigger. Summer new moons are Yang; Summer full moons are Yin.

This process is reversed in the winter. Each new moon takes place with both Sun and Moon in a Southern sign (Libra-Pisces). The initial impetus is less direct, more in the nature of new reflections. But the full moon two weeks later when the Moon is in the North is fiercer, higher in the sky and more outward in its flowering. The full moon illumination or achievement is born of a more inner trigger. Winter new moons are Yin; Winter full moons are Yang.

So in any lunar month, summer or winter, the full moon will always be ruled by a different Wind, will be in a different hemisphere, to the new moon that precedes it.

The Northern signs of Aries to Virgo are sometimes called the *personal* signs, with Libra to Pisces as the *universal*. And again, the Northern signs of Aries to Virgo were once called the *commanding* signs (Ptolemy) while Libra to Pisces were the *obeying* or obedient signs.

Planetary Affinities to the Two Winds
To understand better the meaning of these two Winds, Boreas and Auster and the energies they govern, let's relate the planets to them.

The Sun rules Leo and is exalted in Aries. Both of these are Northern signs so, not surprisingly perhaps, the straightforward daytime Sun is thought to be at its strongest in the time of summer in the North. And the same is true of the Moon which rules Cancer and is exalted in Taurus, both of which are Northern signs. The Moon in the North is a lot more forceful and direct than it is sometimes thought to be.

So far in our examination four out of the total of six Northern signs have been covered by rulerships and exaltations of the Sun and the Moon: Aries, Taurus, Cancer and Leo. The remaining two Northern signs are Gemini and Virgo, which are those that Mercury rules and is traditionally exalted in. So the Sun, the Moon, and Mercury – the first three personal planets – are the planets most associated with the energy of Boreas.

Venus and Mars traditionally rule two signs each, one in the North and one in the South. Venus rules Taurus (in the North) and Libra (in the South). Mars rules Aries (in the North) and traditionally Scorpio (in the South). Although their exaltations are both in the South – Venus in Pisces and Mars in Capricorn – Venus and Mars have energies that can span both Winds, associations with both hemispheres.

And the same is true of Jupiter. While it is the ruler of two Southern signs: Sagittarius and traditionally Pisces, it is exalted in the midsummer sign of Cancer, a high Northern sign. Jupiter is at home with both Winds.

When we come to Saturn and the outer planets we find a totally Southern sign emphasis. Saturn rules Capricorn and traditionally Aquarius and is exalted in Libra. All three are in the South, in Auster's domain. Saturn does not rush in to things.

The outer planets have been accepted to rule southern signs too: Uranus – Aquarius, Neptune – Pisces and Pluto – Scorpio. There is disagreement over outer planet exaltations so we shall ignore them here. It gives further weight to the idea of outer planets being more collective or generational in influence. They are 'universal' and not so personal as the Sun, Moon and Mercury.

Mundane cases
To reiterate our premise – every new moon is a new energy, a sowing-seeds energy, but it is slightly different in summer and winter.

Bombs were the seeds that were sown in London directly following the Cancer new moon of July 6, 2005. England's capital city was in euphoric mood after the news that it had won its bid to host the 2012 Olympic Games, announced almost exactly on the hour of that out-of-bounds new moon. The new moon closely conjuncted by longitude and declination the United Kingdom's natal out-of-bounds Moon in Cancer so it directly affected the people and their mood. The next morning terrorist explosions during the rush hour on the London underground railway and on a tourist bus caused shock and injuries to over three hundred people and a death toll of over fifty. The capital city's transport system was brought to a temporary standstill. This was an outer event. New moon in the North.

Two weeks later on the day of the full moon of July 21, 2005, a shadow of the new moon event occurred when more bombs exploded on London's transport system. This time no one was hurt and although it could be argued that this was also an 'outer' event it was far more muted in its effects. The aim of the perpetrators seemed to be to spread fear, confusion and uncertainty (negative South Wind concepts) amongst the public. All the suicide bombers responsible for the new moon atrocities had been identified by the full moon but within 24 hours (of the full moon) an innocent man was chased by police into the underground and shot dead as he entered a crowded train. This fatal error was the first time anyone had been shot on the London Underground Railway and it happened when the full moon was deeply in the South opposite a Sun that conjuncted and paralleled Saturn. Not only does the negative side of any full moon tend towards illogical emotion-ruled acts but 'muddled identity' is a real South Wind issue. The sense of who am I? is not clear. Name and outer identity belong to the conscious North Wind. The public (moon) reaction to this wrongful shooting also engendered high emotion.

Thus the new moon action had culminated in a Capricorn/Saturn full moon of retribution. Although in the public mind some sort of justice had been achieved by the police in swiftly identifying the original bombers and defusing the second wave of attacks, the full moon brought inner feelings both positive and negative that hinged on the outer event of the new moon.

A year later in the summer of 2006 the Full Moon of 9 August saw a terrorist plot foiled at Heathrow Airport, a plot designed to smuggle explosives on to passenger planes and to trigger them to explode mid-air over the Atlantic. Had the plan succeeded the devastation and loss of life would have been immense. But this disruption was centred on a summer full moon, a Yin energy not a Yang. Although we do not know for certain what the new moon trigger to this event, two weeks before, might have been, the pattern to that of 2005 was similar. The southern full moon defused potential public devastation; it was not active. In terms of air travel to and from Heathrow this full moon in an air sign did engender some cancellations and delays (emotional full moon confusion), but just as in 2005, the authorities made arrests on the full moon, reducing the threat of terrorist action at this highly charged time.

A similar occurrence took place a year later on the full moon of 30 June 2007. Two separate terrorist car bomb attacks, one outside a London club on 29 June and one at Glasgow Airport on 30 June, failed in their objective to cause injury or loss of life. In both cases the general public (the moon) were the heroes of the hour. In London three people noticed something suspicious and alerted the authorities before any bombs could explode. In Glasgow where two suicide bombers drove a flaming car through the main doors of the airport, the plan misfired, one bomber was almost burnt alive, the other dragged out and held by the public. The bomb didn't explode.

This full moon in June 2007 was noticeably strong in relation to the United Kingdom chart. With its Sun and out-of-bounds Moon at 8° Cancer/Capricorn it mirrored the Capricorn/Cancer out-of-bounds full moon on the UK chart and fell across its Midheaven/IC axis of 9° Cancer/Capricorn. It coincided with the start of a public smoking ban in England from midnight that night, the inception of a new Prime Minister (Gordon Brown took over from Tony Blair on June 28th, two days before), record levels of torrential rain causing flood damage to homes and buildings across the country (flooding is a typical full moon phenomenon) and an increased number of full moon road accidents. Typically perhaps for its extra Yin energy a 43-vehicle pile-up in atrocious driving conditions on a Dover road on June 30th resulted in only one death. The accidents increased but the fatalities were less than expected. On the day after the full moon the Princes William and Harry hosted a pop concert to the memory of their mother, Diana (the Moon Queen).

Full moons are always irrational and emotional times but summer full moons, Southern full moons, may be more so.

The Asian Tsunami disaster on December 26, 2004 was on a winter full moon, a North Wind moon. This was a furious headstrong noticeable outer event.

In April 2010 the north wind was literally responsible for blowing a cloud of ash from an erupting Icelandic volcano across the United Kingdom and northern Europe. Although the volcano began spewing smoke at the time of the spring equinox in March it wasn't until within 24 hours of the Aries new moon of April 14th that all airports across the UK were shut down enforcing a total ban on aircraft arriving and departing. Europe's busiest airport, London Heathrow, remained inactive for over five days, unprecedented in civil history. Both the chaos to travellers and cost to the aircraft industry was extensive. This was a summer new moon, the first of the year with sun and moon at 24° 27'Aries. A completely noticeable outer event, no hidden agendas, totally yang. Then a second volcano erupted and the north wind continued to blow more ash into precious European airspace.

Naturally we do not expect to find that every remarkable happening in world history takes place exactly on a new or full moon, but when such situations do coincide with lunations we can detect a subtle difference according to summer or winter.

Moon Princess
Let's now examine a personal life story that has always intrigued astrologers because it clustered around lunations, namely Diana, Princess of Wales.[4] The royal wedding of Prince Charles and Princess Diana took place on July 29, 1981 (11.17am BST London), less than two days before a new moon on July 31, 1981 (03.52am GMT). This new moon was a Northern one at 7 Leo 51, with the Sun and Moon unusually close, as this was an eclipse.[5] Although an eclipse of the Sun is not traditionally a good omen for royalty, especially in Leo, the energy of the day was undeniably an outer, public, noticeable event of show and grandeur.

Diana's first child William, future Heir to the Throne, was born on June 21, 1982 (21.02pm BST London) just hours after a new moon on that solstice day at 29 Gemini 47 (11.52am GMT). This was also a solar eclipse. The northern new moon produced a much-reported outer event, a very public birth.

---

4. For a more detailed and most readable account of astrology and Diana see the 'Moon Queen' chapter in Neil Spencer's *True as the Stars Above*, Victor Gollancz, London, 2000.
5. Coincidentally the House of Windsor itself was founded two days before a northern solar eclipse. George V issued a royal proclamation of the change of family name to 'Windsor' on 17 July 1917 (no time of day was recorded). The solar eclipse two days later was at 25°51' Cancer.

A decade later the marriage of Diana and Charles had more or less fallen apart and the Prime Minister made a public announcement of their separation coinciding with the lunar eclipse on 9 December 1992. This was a northern full moon, a much reported outer acknowledgement of a rumour situation that had never been officially confirmed until that time. The final divorce was announced on August 28, 1996 (10.27am BST London). This was on the day of a full moon at 5 Pisces 41. It was a southern full moon so not such a greatly fanfared outer event, more a quieter culmination that took place away from the public eye without even the participants present. An ending rather than a beginning. We may notice that this was the formal ending to a situation that began on a new moon fifteen years before, so it does not necessarily follow that each full moon brings only to fullness ideas or events that happened two weeks previously. This last full moon was not a lunar eclipse but all the new moons in this story were solar eclipses, and eclipses do have their own distinctive time patterns. However we will stick to our basic premise here.

Diana's death from a car crash in Paris took place a year later close to another solar eclipse. The time and date of her death is recorded as August 31, 1997 at 04.00am. This was one day before a Northern new moon and solar eclipse on September 1, 1997 at 23.52pm GMT, 9 Virgo 34. This northern new moon produced an outer event that would be reported all over the world. Although this was a death, an ending, it was not a private personal culmination; it was a media circus event.

Lastly it is interesting to observe that Prince Charles (whose pre-natal new moon was a solar eclipse) married again on the morning after a northern new moon and solar eclipse that took place on April 8, 2005 at 19 Aries 06. The pomp and ceremony of a Pope's funeral also coincided with this new moon and was probably the more dominant world-interest event.[6] But we can also observe how closely the declinations of the Moon and the Sun at this new moon (07N10, 07N29) parallel those at the death of Diana (07N07, 07N59). The Sun's exaltation at 19+ degrees Aries, reflected in the antiscion around 10 Virgo, when coinciding with a new moon, appears to bring dramatic outer globally reported world events.

---

6. A new Pope was inaugurated two weeks later on a lunar eclipse.

THE GODDESS WITH THREE FACES

'As a celestial divinity she was called Luna, as a terrestrial, Diana, and in the lower world Hecate.'
> [Edward Bulwer-Lytton, referring to the moon goddess in *The Last Days of Pompeii*, 1834]

The Moon and the Number Three

The actual term 'Triple Goddess' for the moon is a comparatively recent one, attributed to Robert Graves in his 1946 book *The White Goddess* in which the moon's faces of new, full and old are recognised as virgin, mother, and crone, from a study of ancient poetry and myth.

But the idea behind this three-fold pattern of birth, maturity and death, and of female influence in groups of three (or three times three) is very old. In antiquity the moon was known by a different name for each of her three shapes: crescent, half-moon and full moon, and moon goddesses like Hecate have been named and depicted with three faces or sides. (Hecate Triodites; Diana Trivia). As it comes down to us today we understand that the young waxing moon is the fertile maid, the full moon is the great mother at the height of her nourishing powers, and the old or dark moon is the strict grandmother, the ageing wise godmother, the hag. The new moon begins; the full moon brings to fruition, the old moon finishes.

So whenever the number Three is mentioned we might suspect that the Triple Goddess is peeping from behind the curtains.

Three Wise Women
Outside of Christianity supernatural beings in groups of three were usually female. Norse mythology had the three Norns and Greek mythology gave us the three Fates, the three Graces and the three Furies.

In Christianity the male God was split into three to form the Holy Trinity – Father, Son and Holy Ghost. Often a difficult concept to grasp, the One God became two men and a genderless 'ghost' or spirit. Although still managing to avoid portraying any of these three figures as unambiguously female, it brought a faint echo of the moon into the sanctity of approved worship. The 'Three Marys' who were present at the crucifixion and resurrection of Jesus, and who were the subject of many historical works of art, are another of the more obvious examples of this poetic three-fold feminine power. We will look a little closer at the Three Marys a few paragraphs on.

The Fates
The Fates or *Moirae* were the three sisters in classical mythology who held the thread of life. Sometimes called the Daughters of Night the daunting

nature of their influence leant them the additional name 'The Three Grey Ones'[1] or the *Graeae*. Identifications may get confused as it was the Graeae who were the grim trio with one eye between them (a right eye no doubt) to whom Perseus was obliged to consult before confronting the gorgon Medusa. The Fates however were more commonly seen as Clotho, Lachesis and Atropos, spinning, measuring out and then cutting the thread of each mortal's life.

The Graces
The three Graces or Charities were the goddesses of the banquet, dance and social play. They encouraged gentleness and beauty.

The Furies
The Furies or *Erinyes* were the classical version of Charlie's Angels, only more hideous. They were the three avenging goddesses who relentlessly pursued those who had committed crimes gone unpunished.

Nine muses
Urania, who rules the study of stars, was one of the nine Muses, and each presided over a particular brand of art, literature or science. Three triads is called an ennead. Besides Urania the other Muses were:

> Calliope, the muse of epic poetry,
> Clio, the muse of history,
> Euterpe, the muse of lyric poetry,
> Melpomene, the muse of tragedy,
> Terpsichore, the muse of dancing,
> Erato, the muse of erotic poetry,
> Polyhymnia, the muse of sacred poetry,
> Thalia, the muse of comedy.

Faith, Hope and Charity
A phrase from the King James Bible that attained great longevity was 'faith, hope and charity' (1 Corinthians 13.13), originally part of an epistle urging the listener to abide by these three qualities. Somehow, because they are three, people unconsciously associated them with the female, and Faith, Hope and Charity have been girls' names ever since.

---

1. According to Robert Graves.

Macbeth's Three Witches

In *Macbeth* 'the Scottish play' (therefore ruled by Cancer and therefore the moon) the three witches who open the first scene are not individually named. While they refer to each other as "sister" they are identified only as Witch 1, Witch 2 and Witch 3, or – as other characters call them – the "weird women" or "weird sisters". The witches speak often of threes and nines, as in this chorus that they chant together in Act 1 Scene 3:

> 'The weird sisters, hand in hand,
> Posters of the sea and land. Thus do go about, about:
> Thrice to thine, and thrice to mine,
> And thrice again, to make up nine:–
> Peace! – the charm's wound up'

The moon goddess Hecate herself appears in the play, though not directly to the mortals. To the witches she refers to herself as "the mistress of your charms" and keeps a strict and wary eye on what the three witches are doing in their spell making. There is no question but that the three sisters look on the moon goddess Hecate as their mentor.

> 'Upon the corner of the moon
> there hangs a vaporous drop profound;
> I'll catch it ere it come to ground…'
> [From Hecate's speech. *Macbeth* Act 3 Scene 5]

For the moon to have a 'corner' it must be crescent shaped, confirming that Hecate is the goddess of the moon's waning crescent phase.

Cerberus

Dogs were especially associated with Hecate – and Artemis and Diana – and mythology gave us a dog with three heads: Cerberus, the guardian to the midnight realm of Hades.

Three Wise Men

The Three Wise Men or astrologer priests who follow the Star of Bethlehem in the biblical story of Christ's Nativity are another interesting case of masculine moon symbolism. The scriptures make no mention of 'three' wise men, simply "wise men from the east" (Gospel according to St Matthew). The number three has been assumed because three gifts of gold, frankincense and myrrh are mentioned. But some unconscious need in the minds of the hearers of this story has made the number of priests Three, has given them a lunar side. These are men both of learning and intuition. They 'follow' a star, they 'react' to the signs they have read of the birth of a new Sun, a new Messiah, a New Age. They do not force events; they appear and then they disappear, with a minimum of outward fuss and show. In the original telling of the story the Wise Men have no specific names. It was only in later centuries that they were assigned personal names and individual identities. Their part in the story is far more moon-like than sun-like.

Nevertheless the Magi's gifts mentioned by St Matthew are all ruled by the sun. Gold is unquestionably ruled by the Sun and both Frankincense and the Myrrh tree are ruled by the Sun according to William Lilly. So a satisfying balance between sun and moon has been achieved in the iconic image of richly dressed kings riding camels at night across sun-ruled deserts. The wisdom comes from the east, the direction of the sunrise and from where the astrologer priests originate; and in this popular nocturnal scene no moon is shown. The Star of Bethlehem must take precedence in the sky and to actually see the moon in this picture would be uncomfortable. Christian iconography has generally preferred to avoid the moon except in certain early representations of the Virgin Mary with the crescent moon beneath her feet or more unusually on her brow, but the intuition of individual artists over the centuries has found it difficult not to paint Mary and the moon together. *The Immaculate Conception* by Velasquez (1618), is a typical example of this with the Holy Virgin standing on the orb of the moon.

It was understandable that Christians were not overtly encouraged to identify Mary directly with the moon as moon worship was regarded as pagan. It may be no accident that the New Testament scriptures retain the story of Jesus prophesying that he will be betrayed three times (before cock crow) by one of his closest disciples. It seems to foster a subtle distrust in the night hours combined with the number three.

While the lunar identification was satisfied subliminally through the maternal images of Mother Mary or of the Madonna and child, the confusion over Mary is curious as one Mary, the mother of Jesus, was worshipped as a virgin saint while the other important Mary in the story, Mary Magdalene, was regarded as a whore. We can sense different sides or phases of the same moon influence. A virgin (new moon) who is also a mother (full moon) deftly combines the energies of new and full moon in the one person of the Virgin Mary/ Mother Mary, leaving Mary Magdalene to complete the lunar pattern by portraying the dark phase.

As mentioned above Three Marys came together to stand beneath the cross at the crucifixion of Jesus, which according to St Luke may have coincided with a solar eclipse ("And the sun was darkened" Luke: 23.45). At a solar eclipse God is supplanted because the moon has blacked out the sun. Other traditions maintain that this was a full moon, in the way that Easter will continue to be calculated,² and that the Crucifixion Eclipse must have been a lunar eclipse.

The three Marys also attended the tomb of Jesus at the resurrection, and apart from naming Mary Magdalene the gospels are a little hazy as to the actual identity of this trio. According to St John's Gospel the Three Marys were Mary his mother, Mary his mother's sister, the wife of Cleophas; and Mary Magdalene. At the resurrection the third Mary is Mary the mother of James according to St Mark and simply referred to as "the other Mary" by St Matthew, or various women from Galilee and not necessarily just three in the Gospel of Luke. It is quite confusing and the individual names of the three Marys clearly do not matter except that they are centred on the figure of Mary Magdalene who is unequivocally identified in this episode in all accounts.

Mary the Virgin Mother was the important moon figure at the birth of Jesus, and Mary Magdalene becomes the important moon figure at his death. For various reasons Mary Magdalene is associated with a dark moon – she is supposedly a fallen woman and certainly not a virgin; she mourns the death of Jesus and will be the first to see him rise. As a dark moon she is a black madonna. The old pronunciation of her name doesn't sound the 'g' and is said as 'maudlin', a word later used as an adjective in English to mean sad, doleful and overly sentimental. These can be seen as waning dark moon emotions.³ Controversy has always raged over her contradictory and unknowable character and persists into the present day with the conjecture that she was no less than the intimate companion or wife of Jesus.

But back to the three wise men…

The Three Wise Men of Gotham

The Three Wise Men of Gotham is an old rhyme built on a legend that the people of Gotham, a village in Nottinghamshire, were all simple. One version has it that the inhabitants of Gotham deliberately cultivated the

---

2. Easter is the first Sunday after the first full moon after the spring equinox.
3. The Oxford and Cambridge Colleges of Saint Mary Magdalene are still pronounced 'Maudlin'.

reputation as fools to divert the unpopular King John and his retinue from building a permanent stopover there in the 13th century.

The name 'Gotham' is better known today as Gotham City the fictional metropolis in which the stories of Batman the Dark Knight are set. Though possibly not consciously intended by its modern creator, this is still a city of fools by implication, rife with crime and corruption. Most of Batman's colourful opponents are psychologically unstable and dress as Fools (the Joker, the Riddler…) while he dons the cape of a bat and operates by moonlight.

To be wise or foolish is always a lunar dilemma. Endless parables about wise men and fools can be found in the popular idioms and almanacs of previous centuries. What may seem mad in one light is not in another. There is often method in madness, and vice versa.

'Three Wise Men of Gotham
Went to sea in a bowl.
If the bowl had been stronger
My tale would be longer'

The short poem is full of lunar imagery (bowls, sea, three). Its appeal stems from the spectacle of sober dignified wise men making fools of themselves. It's the moon (the public, women) laughing at the dry pomposity of those who don't take the moon's influence seriously. It's a theme often met in the art and literature of the late Middle Ages when a figure representing Logic, usually depicted as a scholarly man in sombre robes, is daunted and confused in the presence of Nature, usually depicted as a fair pale-skinned woman.

Only witches could float on the sea in a skimpy craft and survive. Going back for a moment to the witches in *Macbeth*, we find this line:

'But in a sieve I'll thither sail,
And like a rat without a tail,
I'll do, I'll do, I'll do'

It reflects the common belief that witches could float, whether they had a boat or not. For the official witchfinders in those days – or anyone else who wanted to get rid of a woman they didn't like – this was a dangerously useful piece of dogma, for if a woman accused of being a witch sunk at her trial she was dead anyway, and if she floated she had proved herself a witch and could therefore in theory be burnt.

Moon rakers
Another old tale of moons and foolishness is the one about a group of villagers, traditionally from Wiltshire in the English rendering though the theme is universal, who see the reflection of the moon in a pond and try to fish it out. In some versions it is said they believe it is a cheese that had fallen in.

Again this is sometimes backed by a semi-historical homily about smuggled goods being hurriedly stowed in a pond while customs and excise agents are put off the scent by locals pretending to be idiots raking out the moon.

The moon we would assume must be full in this spectacle.

More moon ships: The Ship of Fools
Like the wise men of Gotham in a bowl, the Ship of Fools has much the same lunar imagery. First appearing in Western art and literature in the 15th century *The Ship of Fools* depicts a variant array of men and women happily adrift on a futile sea journey of self-deception.

Hallucination at sea is time honoured. 'Stranger things happen at sea' goes the old expression. Hundreds of miles from land, surrounded by water, a boat is perilously unearthed and subject to the lunar side of reality. Sailors were and still are one of the most superstitious groups of people and the moon, which rules the sea, exacerbates superstition and spirit phenomena.

Show business is another superstitious profession. Each night in a theatre the public, the receptive audience (the moon), may react differently, possibly in accordance with the sign the moon is passing through. Actors and stage performers can never be sure what kind of crowd is in the house and what kind of reception they will get. Nothing is constant. It is the reason why many famous actors still prefer live theatre to the more financially rewarding roles they can attain in film. They can react to the reaction of the audience, and be rewarded with an element of heightened spontaneity and excitement. Symbolically too, actors risk losing their selfhood (sun) by playing many changing parts (moon). Like all artists they are emotionally temperamental, their life swings up and down. The moon is magnified in them.

Lest we think that all ships and boats encourage the worst of the moon, let's not forget the tale of Deucalion or Noah and his Ark positively preserving life on Earth in a floating womb-like home. The biblical Ark drifted for 40 days and 40 nights – a number connected with the completion and fulfilment of the ninth harmonic, or the forty weeks of pregnancy. (If you divide the zodiac circle by 9 the result is 40°). As the square of three, nine is the perfect end result of the moon's influence. [See more on number 9 in the Moon's Magic Square.]

I Saw Three Ships
'I saw Three Ships come sailing in on Christmas Day in the morning' is a Christmas song but its origins are obscure and not all versions make mention of Christmas, nor that the ships are carrying figures from the Christian Nativity. A once popular version still found in Victorian nursery rhyme books had each ship sailed by a woman, and in answer to the line 'Who was in those ships all three?' the response came 'Three pretty girls were in them then… On New Year's Day in the morning'. This is far more moonlike – three women sailing on the sea – and seems more natural than the wording of the Christmas carol describing the Virgin Mary and Christ sailing in three ships to (land-locked) Bethlehem.

The non-religious version ends in a wedding.

Dante's *Inferno* and the Three Ladies Benedight
We meet ladies in groups of three throughout the whole of art and literature and I'd like to mention here just one more famous instance. In the opening of Dante's 14th century epic poem *The Divine Comedy*, the author is a pilgrim who begins his journey entering a dark forest. Today we would probably say he is entering the unconscious realm. As he moves into this shadowy land he immediately encounters three beasts: a panther (leopard), a lion, and a she-wolf. Their wild and daunting nature together with their frightful aspect could represent the ruthless uncanny side of the moon. "*She* [my italics] doth make my veins and pulses tremble". It is the first hazard in this allegorical life-journey thrusting the hero into a place "where the sun is silent".

As the light of day departs a human guide appears; an ancient poet who has been sent by a certain shining lady to help our hero, the traveller. This fair saintly lady is the lovely Beatrice, who comes via the care of the Virgin Mary via Saint Lucia. In other words she is part of a holy female trinity. Lucia – Mary – Beatrice. It is these "Three Ladies Benedight" (three blessed ladies) who wish to oversee the pilgrim's path and help him.

It suggests that the hero's whole experience of the terrors and glories of the journey, and the knowledge and revelation thus gained, is initially the result of meeting and understanding different aspects of the moon.

Any true journey into the unconscious starts with the Moon.

Trilogies

Trilogies in book or film could be seen to be moon-ruled as they come in threes, and it is more usual to find a masculine (Sun or Mars-like) quest or saga taking this form as if to balance itself out. It's uncommon to find a love story written as a trilogy.

The Lord of the Rings is quite a masculine type of book, with plenty of quests, battles and paternal ancestral lines ("son of... son of...") and with female characters in the minority. Yet this is one of the most successful modern trilogies in book and film and many a three-fold saga became fashionable in the later decades of the twentieth century because of it.

J.R.R. Tolkien, the author of *Lord of the Rings*, was none too pleased about his book being divided into three and had great difficulty inventing a title for the middle part.[4] *The Two Towers*, as the second part of the trilogy was finally named, is highly ambiguous. Which two towers are we talking about, asks the reader; there are several to choose from. But Tolkien's publisher made clear the necessity of trisecting the book if the author wanted it published at all. The entire tome was regarded as far too lengthy and expensive to put out as a singular bulky monolith. Tolkien wasn't happy – did he have some trouble with his Moon?[5] – but he had to accept it and the three-part form has become a strong component of the book's identity.

Things come in Threes

Actually they don't necessarily, but if people think they do that's what counts. If it makes us feel more secure to believe that a third disaster or turn of luck will either exorcise or complete a particular run of fate, then there must be a beneficial psychological reason for it. It may seem illogical and unscientific to think that way, but in the province of the moon it could be perfectly the right thing to do.

Perhaps, like a full lunar cycle, experiencing a notion or a happening three times over makes it complete. Carl Jung's Individuation process had three phases: insight, endurance and action. We give people Three Cheers when they have achieved a goal or merit, and Three is a favourite theme in fairy tales where a satisfying end result needs to be won. A common

---

4. See *The Letters of J.R.R. Tolkien*, edited by Humphrey Carpenter, HarperCollins, London, 1995.
5. Tolkien's Moon was in imaginative Pisces inconjunct his 'sensible' Sun-ruler Saturn. J.R.R. Tolkien, 3 January 1892, Bloemfontein, South Africa, 29S09 26E07, late evening.

motif is the presence of three siblings, with the least advantaged eventually outshining the others.

In the story of the *Three Little Pigs* it is the third pig who foils the Big Bad Wolf because he has diligently built his house of bricks, whereas the first two pigs rushed theirs through with sticks and straw. The obvious moral is that successful completion takes time, like a full lunar cycle. In almost every fairy tale something has to happen three times before the bad-wolf/wicked-stepmother/evil-giant is thwarted, before the treasure is found or the prince/princess's hand is attained. Jack and the Beanstalk had to steal three gifts from the giant's castle before he could live happily ever after – a bag of gold, a hen that laid golden eggs, and finally a magic harp; a progression from money to self-perpetuating money to life's higher values of art and music. Cinderella, who had two older step-sisters, attended the prince's ball three times (in the original story) escaping home on the first two nights before her magnificent dresses turned back to rags. On the third night she lost one of her shoes in the hurry to get away and this 'accident' gave the prince a tangible means of finding her, and eventually marrying her. It is as if the scene needed to replay three times before it could become real.

And the fairy tale list is never-ending: Beauty, the third daughter of a bankrupted father, who must meet and transform the Beast; the princess who had to take the frog to bed three times before he turned into a prince; Snow White's three visits from the malevolent queen and three birds that visited her glass coffin before a prince could find her; three wishes; three feathers; three spinners, three heads in a well … and not forgetting Goldilocks and her Three Bears – golden sun meets family moon.

There are lunar themes to be found in all these old bedtime stories and it's no surprise that the vast majority of classic fairy tales are told from the point of view of the developing female; although they continue to speak hidden wisdom to both sexes.

The New Moon

'Into itself did the eternal pearl receive us, even as water doth receive a ray of light, remaining still unbroken'
 [Entering the Sphere of the Moon. Dante Alighieri, 1265-1321.
 The Divine Comedy: Paradiso. Longfellow translation.]

The New Moon

This vital point of new birth and beginning is also the darkest possible night of the lunar month. The moment of new moon in astrology, when the sun and moon are together in exactly the same degree of longitude in the sky, allows no sliver of Luna to be seen.

For our purposes, as astrologers, this darkness is the moment of new moon and its position by sign and degree on a chart indicates the area renewed and reborn for the following lunar month. If it falls closely on a natal planet, that planet should be activated in the month ahead. But if the new moon falls, as it usually will, in an empty part of the chart, it is that house area, that earthly department of life, which becomes the background focus for the next 29/30 days.

An exact conjunction of a new moon to a natal planet is comparatively rare and would only happen once every 19 years.[6] One of the most important of all is a new moon on the degree of a natal Sun; in other words a new moon on your birthday. This puts a new moon on the Solar Return chart for that year and suggests that the whole year is likely to incorporate change and new beginnings.

The new moon is birth, but it is birth in the dark. The road ahead may not be lit but you should nevertheless contemplate the first steps on that path. The visible sign of the moon's first narrow crescent may, in northerly latitudes, be two nights away. Nearer the equator it is possible to see the thin bow in 24 hours, but either way the moon's purpose doesn't make itself clear at once. Some see the astrological moment of new moon as conception, with the following visible crescent as the outer birth. In earlier times astrologers took the first evening appearance of the crescent as the basis of a chart to predict the coming month, or of the entire coming year if this was the first new moon crescent after the spring equinox.

By the time of this growing crescent, illuminated of course directly by the sun, it is sometimes possible to see the rest of the moon glowing faintly

---

6. That is, it would repeat within one degree (60 minutes) of the same place every 19 years.

too. The phenomenon is called 'earthshine', where the ashen glow of the sun's light reflects back on to the moon from the earth. One of its most picturesque names is 'the old moon in the new moon's arms' which resonates well with the astrological meaning of completed or uncompleted business of the previous moon now under a new impetus or command.

In other words the tides have turned.

Any phase of the waxing moon is conducive to growth with the new moon especially generating a lower number of positive ions at the earth's surface, which is said to increase vigour. The span of time between the actual moment of new moon and its noticeable first thin crescent is ideal for making a new intention or beginning a new project, venture or business – even though you are operating by instinct, in the dark. This may be why asking a horary question during this interlunar time would produce a chart with the Moon combust and therefore not strong by those rules. Asking questions implies doubt and the need for concrete answers and this is not a time for doubt but blind faith. Because the new moon produces the darkest of nights, it has sometimes been seen as an inauspicious moon and not recommended for action. But our trust in the growing moon to increase from this moment is based on our memory that she will do what she has done many times before. The growth has begun and we should trust this habitual pattern, this promise of a breath of fresh air, rather than the logical evidence of our own eyes, which sees only a moonless night.

The New Moon is the poetic virgin moon, the untamed maid who is at one with the wildness of nature; the young princess whose potential lies before her. The new moon is that princess, innocent and naïve in some ways yet already a little woman carrying the untaught knowledge of future brightness and darkness. New ideas that are instinctively birthed at new moon can wax and grow more rapidly and outwardly. Ideally every new moon should be used to symbolically sow seeds or renew oneself in this spontaneous way.

We can also say that new moons run in seasons. That is, the zodiac degree that the new moon tenants in any one month is likely to be the same degree in the following month, though obviously one sign further on. For instance, if you look in an ephemeris, there might be a run of four or five months when each new moon is at 29 degrees of successive zodiac signs – an end-degree energy that could denote a season of summations, of accepting and completing old business; albeit with the impetus of new moon drive. In a similar way a run of new moons at zero degrees of the signs, at the very start of the signs, suggests a season of lively new beginnings. Wherever the new moon degree is placed from 1° to 29° it might be in exactly the same degree as one of your natal planets and will therefore continue to dynamise that part of your chart in various ways and varying aspects (sextile, square, trine, inconjunct…) as it makes its lunations in different signs over a period of several months. The same thing applies in a more universal way if there is a slow-planet conjunction in the sky in a particular degree and if a run of new moons takes place on a similar degree. In that case the new moon 'season' would continuously heighten and vary the effects of the longer standing planetary conjunction.

Some NEW MOON charts
The Paparazzi
If each decade of the twentieth century had a city that exemplified it, Rome is often quoted as the place to be in the 1950s when film stars and fashionable celebrities flocked to Rome to mingle with real aristocracy.

Paparazzo was the name of a photographer in Federico Fellini's film *La Dolce Vita*. The name, which is said to be similar in sound to a word for mosquito, could imply a hovering parasite. It caught the imagination in the plural term *paparazzi* as an apt description for the nuisance-making photojournalists that had begun to emerge in society at that time.

According to Francesca Taroni, in her book *Paparazzi The Early Years*,[7] its defining incident occurred at 2am on Saturday 16 August 1958, at the Café de Paris on Rome's Via Veneto, when a pushy freelance reporter leaned in a little too close to photograph some ladies' legs and was hurled back on to the street by the party's host, Egypt's King Farouk. Legend has it that a blurry shot of the fracas survived. Later that night two film stars, who shouldn't have been together, were spotted having an emotional argument at a nearby nightclub and the photographers rushed over in a small pack on the scent of more scandalous images. Once again the outraged male victim physically lashed out at the pests…

Such intrusive pictures soon found a lucrative international market that hasn't abated. But the new trick that started in Rome was to be at the right place at the right time and if necessary create news rather than simply recording it.

On the chart the Sun is in Leo, the sign of theatrical stars and celebrity, and the ruler of Rome. The new moon is in Mercury-ruled Virgo in the

---
7. Francesca Taroni, *Paparazzi The Early Years*, Assouline, Paris, 1998.

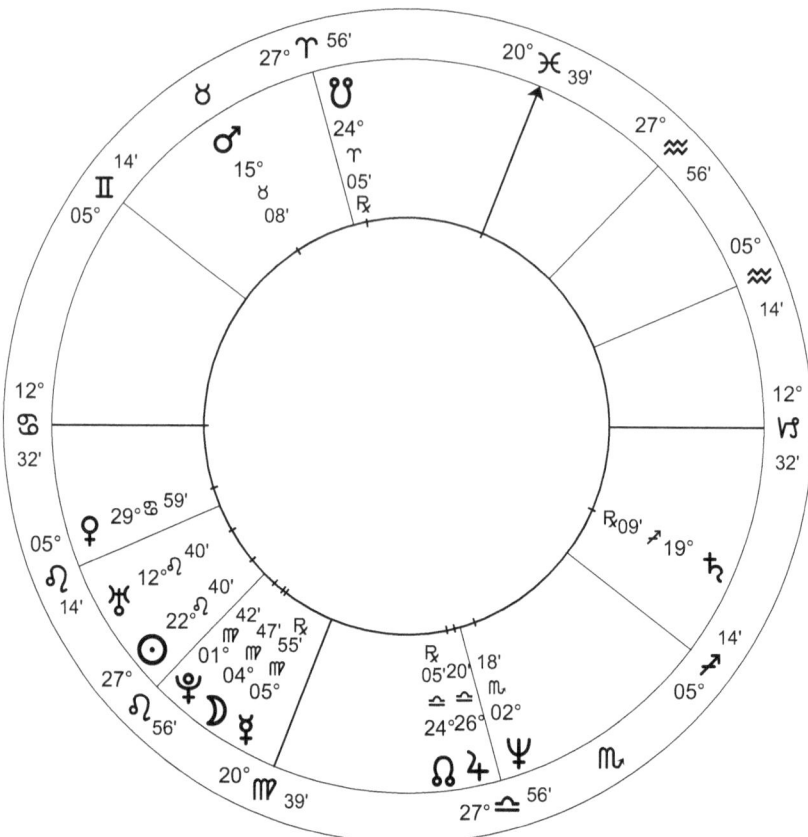

The Paparazzi: 16 August 1958, 2am, Rome, Italy. Source: see footnote 7.

third house and conjuncts Mercury and Pluto. Media and journalism belong to Mercury and the third house, and here was the beginning of a trend away from the polite respect and consideration once afforded to those who were famous towards the no-holds-barred scramble to photograph popular public figures in any light, the more embarrassing the better. The Moon is chart-ruler, happily sextiled to Neptune, suggesting that reality and responsibility were not over-riding concerns. A new era had begun.

On a more sombre note, many decades later it was the paparazzi who were blamed indirectly for the death of Princess Diana in Paris on 31 August 1997. Her car was speeding to out-race the media swarm when it crashed in a tunnel. It happened on the day before a new moon.

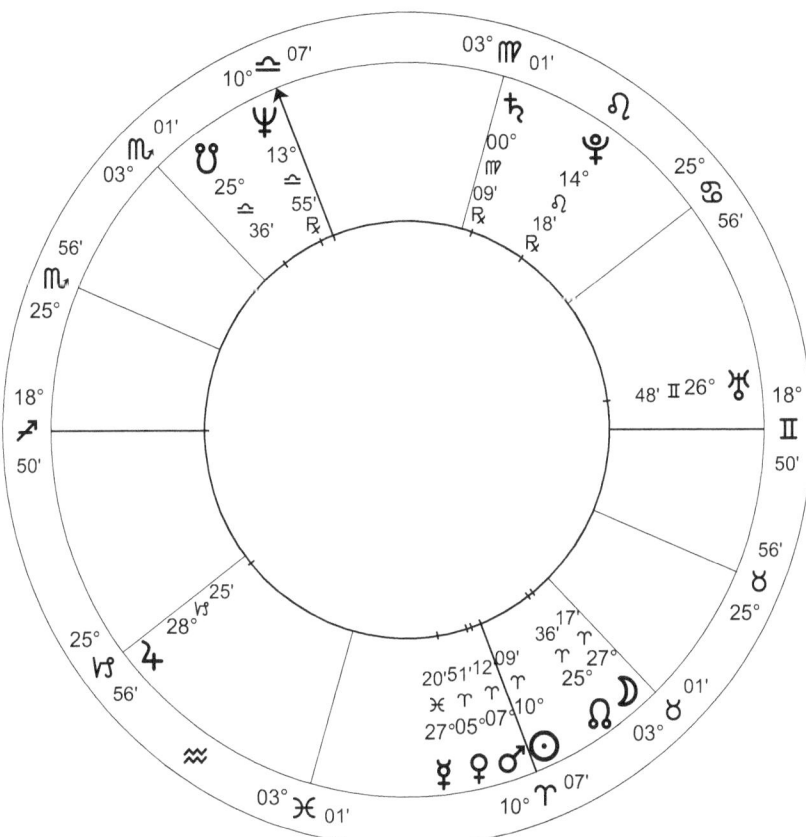

First 45rpm record: 31 March 1949, 00.00hours EST, New York, NY.

The first 45rpm record
The world's first 45rpm records went on sale Thursday 31 March 1949, two days after a new moon in Aries. The resultant chart, taken symbolically for New York, although the records were on sale all across the United States on that date, has a trailblazing Aries energy with the Sun, Mars, Venus, North Node and the Moon all in Aries. Musical Neptune is on the midheaven and expansive chart-ruling Jupiter is in the second house of money.

Perhaps the dreamy glamour of an indefinable artistic ideal (Neptune opposite Sun) made the 45rpm record a popular collectable item, not only during its forty-year run but retrospectively ever since. The earlier large 78rpm records and the later musical cassettes never became collectors' favourites in the same way and are subsequently worth much less in monetary terms. For the 45 the musical reverie of Sun-Neptune was earthed by a Sun trine Pluto and a Moon trine Saturn supplying melodious inspiration in the form of practical, upgradeable, hardy, reliable tactile objects.

Originally referred to as doughnut discs because they had large holes in their centres to make multiple disc playing easier on the new automatic record players, there was not one specific title that can lay claim to be the first 7-inch 45rpm disc ever released. Records by 104 different artists were issued simultaneously by the RCA Victor record company on that day of 31 March 1949 and every one of these had previously been released by RCA on the old-fashioned 78rpm records. (It may be one reason why the Sun and Mars can be seen hovering around the IC). Nevertheless the 45 was destined to make its name as a cultural artefact, particularly in pop and rock music, changing the musical landscape forever as it did so.

Darwin's Origin of Species
Charles Darwin had delayed publication of his *On the Origin of Species by Means of Natural Selection* for many years for fear of the public furore it might cause.

To the amazement of its publishers, who had tried to persuade the author to put out a safer work on pigeons, Darwin's book sold out on its first day and became the most widely read scientific text in world history, widening our views, changing our understanding of our origins fundamentally and challenging dogmatic religious beliefs, (the new moon is in Sagittarius in the 9th opposite Uranus in the third). The symbolism speaks for itself, and although Darwin tried to emphasise that God was still the source of nature, for many people this new idea was totally shocking, blasting long-held securities at a stroke. Charles Darwin, the often physically indisposed student of botany and natural history; Charles Darwin the happy family man who never wanted to upset his friends; Charles Darwin who put forward his ideas in a tentative way always allowing that he may not be correct on every point, was branded the most dangerous man in England. Like a new moon *The Origin of Species* had really started something.

This chart is set for the exact moment of the new moon and as in any precise new moon chart the Part or Lot of Fortune will be found situated on the ascendant. It proves again the vitality of birth at a new moon, that the part of fortune or luck meets the world directly, earthing the power of the moon recharged. The Part of Fortune is also known as the 'Ascendant of the Moon' and at new moon the title is literal.

At the time of this new moon in London the sensitive and unscientific Neptune in Pisces was rising, conjunct the ascendant and the Lot of Fortune,

The Moon and Number Three   37

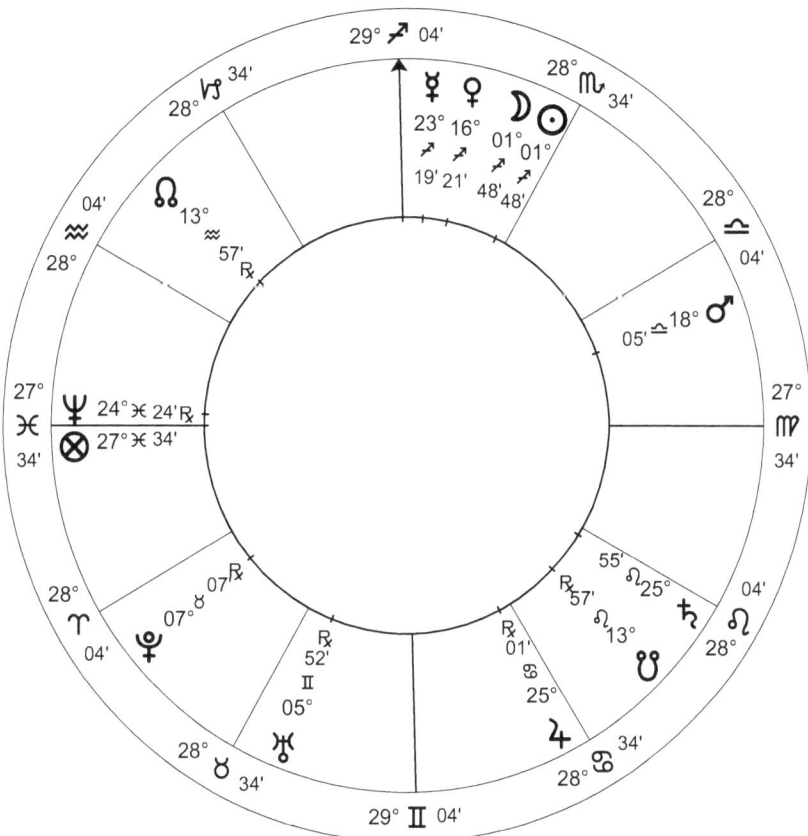

Origin of Species: 24 November 1859, London, England. Time set for moment of new moon on that day 1.44pm

suggesting a direct erosion in faith and belief caused by Darwin's bombshell and highlighting the shaky ground many people's beliefs now stood on. The only way to deny Darwin's discoveries was to close your ears and pretend he would go away, or to reinforce your conviction in what you were taught as a child. (Namely that all animals and birds were created in one day and they had not altered in any way from that moment). In either case the response could be a Neptunian one, an escape from reality. On the other side of Neptune's coin, and another meaning of Neptune rising conjunct the Lot of Fortune, is the inspiration it gave to countless others to adapt this knowledge to their own feelings and beliefs. This in turn had its negative as well as positive outcomes and in due course the process of natural selection, or survival of the fittest, was seized on by some to champion tenets of either capitalism or communism in equal measure. A couple of generations later its misinterpretation met its darkest hour in the racist doctrine of Nazism.

Neptune was discovered in 1846, only a decade or so before this chart and true to the planet's nature its discovery was confusing. Unlike the sudden sighting of Uranus by Herschel in 1781, several different names can be connected to the hunt for Neptune and typically its position and orbit took some time to pin down. In a similar way Darwin's theories on natural selection had been formulated years before the general public came to know about them and had also been arrived at by the naturalist and explorer Alfred Wallace[8] quite independently. The two men welcomed the collaboration of each other and it was the knowledge of Wallace's discoveries that persuaded Darwin finally to publish his own. He acknowledges "Mr Wallace's excellent memoir" in the Introduction to his book, confirming their similar viewpoints. At this time Darwin was in England while Wallace was in the Malay Archipelago.

Another facet of this chart's ninth house emphasis is Darwin's admission that his own theories were originally formed when he was travelling on a sea voyage far from home. *The Origin of Species* opens with these words: "When on board HMS *Beagle* as a naturalist, I was much struck with certain facts in the distribution of the inhabitants of South America…"

Charles Darwin, like Alfred Wallace, was born under an old rather than a new moon,[9] hence perhaps his reluctance to rush any new theories out into the public domain. But *The Origin of Species*' new moon at 1°48' Sagittarius was conjunct Darwin's natal Saturn. When he did publish it had the weight of many years testing and refinement behind it.

The Crescent Moon Symbol

As a shape, a crescent is a female symbol. It is cup or bowl-like, a container, a receptacle. According to latitude and season the crescent moon does sometimes appear as a bowl with its horns pointing upwards, although it is more commonly seen and illustrated with the horns pointing left or right.

Drawing the moon in the shape of a crescent identifies it specifically as the moon. Drawing it as a round circle could confuse it with the sun so the crescent symbol is not always being employed to relate to just that particular lunar phase.

---

8. Born 6 January 1823, Usk, Wales.
9. Charles Darwin, 12 February 1809, 3am, Shrewsbury, England. Source: *Astrological Association Newsletter*, May 1995. Other sources have speculated different ascendants.

Neither is the crescent symbol an accurate reproduction of how we actually see the crescent moon in the sky. The illustrator's task has always been to enhance the picture and give a pleasing representation with exaggerated horns and curve. As an emblem a crescent moon is one of the hardest symbols to draw freehand, much more difficult than a star or a sun. Drawing a crescent moon can be a frustrating exercise; it reminds you how nonsensical and unreasonable the moon can be. Try it yourself with a couple of coins. [Then see Appendix: 'How to Draw a Crescent Moon']

A veiled woman
The crescent as an emblem stretches back into the earliest of times. A crescent moon was the symbol for instance of the city of Byzantium, later known as Constantinople and Istanbul. It is said that the Turks who conquered the city continued to use its flag, which led to the Ottoman Empire and the Muslim world adopting the crescent moon as a symbol. The Turkish flag with its crescent moon (and star) on a red ground is one of the oldest surviving national flags and coins with similar symbols date far back into history. Legend has it that Osman, founder of the Ottoman Empire, had a powerful dream of a moon that grew into a tree and covered the earth; the crescent moon bridging from horizon to horizon. The crescent symbol in western heraldry dates from the meeting of European knights with Middle Eastern culture in the era of the Crusades. In Britain it appears on arms in the reign of Henry III (1216-1272) onward, and this design usually had the horns pointing upward.

The curved sword of the Saracen warrior, the scimitar, is similarly crescent-shaped. *Islam* means 'submission' (to the will of God); by analogy acting like the moon to the sun. Many Islamic countries depict the crescent moon on their national flags and this moon is almost always drawn with its horns facing to the right, which is a waning crescent. In astrology we draw the Moon symbol on a horoscope with the horns pointing to the left, which is a waxing new moon.[10]

An explanation for the use of the waning crescent on the red national flag of Turkey is that it was derived from the reflection of the moon in a pool of blood on a battleground (opinions differ as to which actual battle). This means that the moon in question, the moon in the sky, was a waxing moon at that moment; a young crescent moon and not a waning one. Its image in

---

10. At least it is in the *northern* hemisphere. In the southern hemisphere the moon's phases are like a mirror image.

this powerful fable was reversed to the observer looking down on to the pool of blood. The right-pointing waning crescents found on Islamic flags and architecture are therefore new moons not old ones and this accords with the fact that all Islamic months begin with the visual appearance of the new moon crescent.[11]

Taking another instance of crescent emblems, the hammer and sickle flag of Soviet Russia had a crescent moon. In this case intended to be the instrument of the agricultural worker (the sickle) wedded to the hammer of the industrial worker, the appearance of that yellow crescent on a red ground looks very familiar in lunar terms. It was a left-facing crescent too – definitely a new moon. Karl Marx, generally regarded as the father of modern communism, was born on the day of a new moon.[12]

This red flag can engender hot emotion (moon feelings), either for or against its ideals, and its display is still legally banned in some of the countries once occupied by the Soviet Union.

Horns and bow
In various cultures the horns of the crescent moon have led to its identification with horned animals, notably the bull and cow of earthy Taurus in which sign the moon is exalted. Mythical horned half-human creatures like fauns, satyrs or the great god Pan are therefore under the moon's dominion. Pan in particular, from whom we get the word 'panic', is a representation of the wild unreasonable energy of nature. Rustic habits have a close association with the moon. Pan preferred to sleep in the hottest hours of the day, hiding away from the sun's light, and although he can be seen on the one hand as a masculine phallic symbol his earthy fecundity aligned with his irrational behaviour make him more moon-ruled than sun-ruled.

---

11. *The Book of World Horoscopes* by Nicholas Campion shows a chart for the beginning of the Muslim era – Mohammed's arrival in Medina at sunset 16 July 622 AD. On this chart the Moon is about thirty degrees forward of the Sun – a new moon crescent – which accords with the tradition of the visible new moon at the Prophet's flight to Medina from Mecca. Ken Gillman writing in *Considerations*, Mount Kisco, New York, August 2002, makes a persuasive case for the crescent to have been visible on the evening before; that is 15 July 622, and in a more astrologically significant position – conjunct Venus (moon and star) and conjunct Regulus – suggesting that this better signifies the Hegira, the start of the Muslim calendar.
12. Karl Marx, 5 May 1818, 2am, Trier, Germany. Source: Birth records of Trier, quoted in Robert Hand's article on Whole Sign Houses in *The Mountain Astrologer*, Cedar Ridge, California, August, 1999.

Anyone who has read the novel or seen on film C.S. Lewis' *The Lion the Witch and the Wardrobe* may be intrigued to work out into which camp some of the inhabitants of the magic land of Narnia are placed. Do they come under the Sun or the Moon? Everyone in that parallel universe appears to belong to one of two opposing sides. They are either in the domain of the lion king Aslan, golden masculine representation of the summer Sun, or the cold and frosty night-time witch of winter, female representation of the Moon.

The story abounds with ancient themes. The Lion as the sun king dies and is reborn on a stone table, shown (in the film) as an altar in an ancient solar temple. The Witch, with the moon riding high in the sky behind her, delivers the fatal knife thrust that kills him. The Witch's main henchmen are wolves, animals that astrology has always identified as lunar, yet the horned fauns and satyrs are not necessarily on her side. Probably because Lewis was a Sagittarian himself he unconsciously placed centaurs on the winning side of the Sun king and by association all half-animal-half-human mixtures followed.

Archery is aligned with Sagittarius, whose name derives from 'arrow'. But the bow that shoots the arrow has long been associated with the moon because of its crescent shape. Artemis or Diana, the classical moon goddess, is always depicted as a huntress with a curved bow.

Lunisequa – The Lunar Star

> 'Some I have heard say, and others write, that there is a starre which never separateth it self from the moone but a small distance; which is of all stares the most beneficiall to man. For where this starre entreth with the moone, it maketh voyde her hurtfull enfluence, and where not, it is most perilous.'
>
> > [From *The Observations of Sir Richard Hawkins in his Voyage into The South Sea in the Year 1593*. Published in London 1622.]

Crescent moon designs often include a solitary star nestled within the horns of the moon, an astronomical impossibility because the unlit part of the moon would block out any star in this position. But a long-standing poetic belief not only encouraged the idea of a special star that clung closely to the moon but even gave it a name: 'Lunisequa'.

It is mentioned in Thomas Lodge's Elizabethan drama *Rosalynde* in 1592 as if it were a commonly known fact, ("…for as the moone never goes without the starre Lunisequa, so a lover never goeth without the unrest of his thoughts"). And it still appears centuries later in the poetry of Rupert Brooke in a piece written c.1910, ("…and be like the star Lunisequa, steadfastly following the round clear orb of her delight…")[13]

The early descriptions of Lunisequa do not necessarily imply that it is seen within the moon's horns and could therefore refer to that fairly common sight of a bright planet situated close to the lunar crescent. Most usually this would be Venus, and so the Moon would be conjunct Venus, not surprisingly lending it the "beneficiall" qualities that Richard Hawkins described above. It makes void the Moon's hurtful influence, he says if we paraphrase his words into modern English, though he adds it is "perilous" if it isn't conjunct the Moon. Venus separated from the Moon can hardly be described as perilous, so we must take Lunisequa as a special phenomenon. Any planet or star temporarily sitting next to the moon is 'Lunisequa', whatever planet it may be, even though we know it has a separate identification elsewhere. And when it is 'Lunisequa' we should be filled with delight.

---

13. Rupert Brooke, 1887-1995. From *Thoughts on the Shape of the Human Body*.

The Full Moon

'The moon is ladye of moisture'
[Isaac Newton 1574]

The Full Moon

At full moon the moon is at its totality. No shadow from the Earth blocks it and it is fully illuminated by the sun's light. This is the moon at its greatest strength. It is full and complete and life is full and rich. Poetically this is the fertile Mother, the moon at her most moist, maternal, blossoming, sustaining, fruit-bearing, life-giving, nourishing and in every way abundant.

In the psychology of Jung, symbols connected to the mother archetype read like a list of things ruled by the moon (sea, hare, cow, cave, font, oven, witch…). While the psychological list ranges even further to encompass many items and states, one of the most telling of Jung's statements names the basic aspects of the mother archetype as threefold: 'her cherishing and nourishing goodness, her orgiastic emotionality, and her Stygian depths'.[14]

Positive ions at the earth's surface are said to increase at the full moon, which is regarded as a negative effect or at least an unstable one, on nature and living creatures. The atmosphere seems 'dusty' and stormy. There is tension between the opposing spheres of moon and sun. This can manifest on earth in the realm of human relationships, on our actions and reactions to others. You may be taking extra care if you are driving at full moon for example, as indeed you should, but it's the others you've got to watch.

And relationships in general are very important to those born at full moon at all times, not only because the opposition is a relationship aspect – and this one involves the two planets of paramount significance – but also because at full moon the Part of Fortune will be conjunct the descendant.

Modern science continues to discover reasons for the full moon's specialness – that the moon is noticeably charged as it travels through the Earth's magnetotail at this time for instance.[15] Anecdotally it is the full moon that has the greatest pull on the human condition ranging from domestic argument and violence in the home, to road accidents or near misses, to mental instability both joyous and dangerous, to irrational, erratic, disorganised, suicidal, loony and howling behaviour of all kinds. In a gentler

---

14. C.G. Jung *Four Archetypes*, Routledge and Kegan Paul, London, 1972. Originally published as a lecture in 1938.
15. The area fanned out behind the Earth where the pressure of the solar wind has collided with the Earth's magnetic sphere.

way a full moon can cause people to 'moon about' and happily go about their business in a zombie-like fashion, seemingly somewhere else in mind and spirit. (Try visiting a supermarket at full moon). And inner illumination can suffuse the thoughts and feelings with long-sought understandings or arouse half-forgotten memories of incidents and people from many years past, especially if they were emotionally charged experiences.

The moon's pull on water, of which the surface of the earth is predominantly composed, inclines any liquids to flow more freely at this time. At full moon the tides run high up the beach and rivers flood. The breaking of the waters in pregnant women is more likely at full moon too, as is the need to avoid surgery as the blood flow is more difficult to stem.

On a cloudless night the full moon lights the landscape with a power nine times that of a half-moon. It offers the brightest natural illumination possible in the darkness of night. And a full moon will be larger in the sky when it coincides with its perigee, the time that the moon is closest to the Earth in its monthly orbit.

A *rising* full moon, low on the horizon, will also appear dramatically large, and this only happens at sunset or soon after. The sun is directly opposite the moon and as the golden god sets into the land or sea in front of us the silver goddess silently arises immediately behind us – and in her most magnificent form.

Special Full Moons

Other than lunar eclipses we don't usually distinguish between the different full moons that take place during a year, except that they will be in different zodiac signs or perhaps fall on a degree already charged by transit or have particular aspects to them. But some are celebrated annually or are special for various reasons and we take a look at a few of these below.

The Harvest Moon

Popularly, rather than astrologically, each full moon has been given a title like the Hunters Moon (in October) and most famously the Harvest Moon in September. The Harvest Moon, once sacred to Diana, is the full moon nearest the Autumn Equinox and that does give it some interesting features.

The calendar date of the Harvest Moon will differ but it will be either when the Sun is in Virgo and the Moon in Pisces, or when the Sun is in Libra and the Moon in Aries. This means that it lies close to the equinoctial line – the line of zero degrees of declination – the line that divides the zodiac and our Earth into the northern and southern hemispheres. Lunations here cause the highest sea tides by the way.

The zodiac signs of Virgo, Libra, Pisces and Aries are those that border the equinox points or the celestial equator. So although the Sun and Moon are opposed at full moon by longitude, as they are at every full moon, in this special case they are not very far apart by latitude or declination meaning they are almost parallel and therefore acting like a conjunction. They are opposite by one set of co-ordinates and conjunct by another. It makes it a rather unusual full moon.

The Harvest Moon rises at sunset with a lesser gap of darkness than is the case with moons at other times of year, giving light to help the gathering in of the harvest. When it rides low in the sky it appears physically huge. An Amazon moon. This is a trick of the light, naturally the moon doesn't actually alter its size, but it's a good analogy for the meaning of any rising planets: they are magnified in influence.

The Wesak Moon
The full moon each year when the Sun is in Taurus and the Moon in Scorpio (usually in May, sometimes late April) is celebrated as the Buddha's birthday or the Buddha's blessing and is a focus for planetary enlightenment. It is a public holiday in many countries.

While any full moon offers illumination, literally and symbolically, this one, which is associated with the Buddha's birth, his enlightenment and his passing, is honoured annually as the time he returns to bless the Earth.

As the date and the zodiacal degree of this moon will vary from year to year, there will be some years when the Great Invocation would have a greater effect on your chart than others depending on the aspects it makes to your own planets.

Some years there may be two full moons when the Sun is in Taurus, one at the beginning and one at the end of that zodiac sign. There appears to be no unanimous decision as to which of these represents Wesak. Some spiritual organisations and countries favour the first lunation, some the second. In the West if the second is favoured it may be because the solar month is then May. But the first is often preferred because it would be the second of the (usually) three full moons falling between the spring equinox and the summer solstice. All three of these full moons, and their new moons, are regarded as special, probably because they are part of the waxing, growing energy that is felt all across the northern hemisphere of the Earth at this time. The land is fertile, days are lengthening, sap is rising, everything's on the move.

The only surviving western festival that honours both Sun and Moon takes place within it – Easter. And the once important festival of Pentecost, 50 days or 7 weeks after Easter always falls in this three-month period.

The fact that the moon is technically in its fall in Scorpio at the Wesak full moon appears to be outweighed by these other considerations. It may lead us to deduce that the Moon in Scorpio or Capricorn (its detriment) is not so inexpedient when it is a full moon or if the sun is in the opposite sign. When the Sun is in Taurus it is occupying the place of the Moon's exaltation, and when the Sun is in Cancer it is occupying the place of the Moon's rulership, so the Moon gets a natural boost wherever it might be placed at that time. If the Moon is placed directly opposite the Sun (full moon in Scorpio or Capricorn) it not only rules or exalts the Sun but makes an exact aspect to it. The full moon is now illuminating the Sun in its two most moon-friendly signs Taurus and Cancer.

Paschal (Easter) Moons
There is more to be said about the first lunations in Aries that coincide with Easter.

Easter is based on the first full moon after the spring equinox. This Paschal full moon will always have the Sun in north declination and the Moon in the south, but the new moon that precedes it may have been either a Northern new moon or a Southern one.

In a year when Easter comes early, a full moon might take place on 21st March or very soon after the Sun has entered Aries at the spring equinox. The new moon preceding this would have been in Pisces (in the south) not in Aries (in the north). In a different year with a later Easter, the full moon could follow a new moon in Aries. In other words the kick-off point of its energy would be from a different hemisphere. This surely would colour the meaning of Easter differently each year.

Bewildering Lunations
A Blue Moon
This term is understood to mean different things. While the expression 'once in a blue moon' means that the timing is rare, some understand it to refer to the second full moon in a calendar month or again the second full moon in the same zodiac sign. (Neither of which are exceptionally rare). It

might in addition refer to an extra full moon in a season, meaning the three-month period between solstices and equinoxes.

The implication that a blue moon is something unusual may stem from the fact that the moon hardly ever appears to be blue in colour. White, yellow, or even red; but never blue. Symbolically in hermetic heraldry the moon is depicted as silver, red or black – but again never blue.

New or Full Moons in the later degrees of a sign
When a full moon (or a new moon) takes place towards the end of a zodiac sign, for example 25° Aries, there is a fair chance that its conjunction or opposition with the sun will be the last aspect it makes before leaving that sign. It will then be void-of-course (in the modern understanding of the term)[16] for several hours immediately following the lunation.

These full or new moons may be more ambiguous in their initial impact, unnerving, uncertain and indefinable.

This naturally depends always on the positions of other planets, for if slower-moving planets are in the latter degrees of signs the moon may aspect them after the lunation and not be immediately void of course. But apart from that consideration lunations in later degrees will follow in runs of several months when all the new moons or alternatively all the full moons occur at roughly the same degrees of their signs.

We may then experience a season of void-of-course lunations, when each new or full moon is followed immediately by the aimless, lacking in purpose energy of the void-of-course phenomena. As a rough guide as to how long this empty period might linger, multiply the remaining degrees in a sign by two and call it hours. A lunation at 25° Aries that will become void-of-course will be five degrees from the end of the sign; so 5 x 2 hours (ten hours) will be an approximation of the duration of the void-of-course period.

As with new moons, full moons run in these seasons of several months tenanting the same zodiac degree in successive signs. Wherever the full moon degree is placed from 1° to 29° it may be in exactly the same degree – though not necessarily the same sign – as one of your natal planets and will therefore continue to illuminate this part of your chart in various ways and aspects (sextile, square, trine, inconjunct...) over a period of several

---

16. According to the discoveries of *Project Hindsight* the Moon was only regarded as void-of-course by the Ancient Greeks when it had not been in aspect to another planet for the length of a day or night. Changing signs didn't come into it.

months. The same thing applies in a more universal way if there is a slow-planet conjunction in the sky in a particular degree and if a run of full moons takes place on a similar degree. In that case the full moon 'season' would continuously attempt to illuminate various meanings of the planetary conjunction.

The Moon's a Crowd

Three's a crowd and the Moon rules crowds. The Moon rules the public in general too, the masses, especially if you are comparing them to a ruling government or elite. The Moon in the tenth house can bring its bearer a public popularity.

When individual people combine to become a crowd strange emotional forces take over. Many people attend football matches or large pop concerts as much for the feeling of losing their identity into a greater sea of feeling as for the show itself. And crowds on the streets, as police and law-keepers are ever aware, can turn in a matter of moments from peaceful citizens into a baying mob.

Collective values and people *en masse* are ruled by the moon; and that not only includes protestors in a mob demonstration but also the ranks of authority figures, soldiers or police, who are lined up against them. These are, or should be, the hand of Saturn; keeping to orders, acting within strict guidelines, not going berserk or succumbing to the emotional heat and disorder of the moment. But as countless pieces of media footage show us, order often breaks down and individual soldiers or law-upholders can get just as caught up in the overpowering emotion. Acting in a rogue manner they may mirror the worst excesses of troublemakers in the crowd. If it happens to be a full moon everything is exacerbated; orders may get misinterpreted, intentions on both sides confused and the whole situation incendiary.

Riots have broken out in every era and every country in the history of the world. They do not always happen on a full moon. With the exception of those people directly involved most public demonstrations or riots are soon forgotten by the majority of citizens but occasionally some angry and bloody protests come to be seen as historical turning points and are kept alive in common memory. As in wars and battles it is by no means the number of casualties that decides their importance, it is the effect it has on the rest of the people of that country and/or the world. Cold statistics are secondary to the transmission of human feelings.

Some FULL MOON charts
The Discovery of DNA

No one knows for certain the actual time to the minute that Francis Crick and James Watson excitedly announced their discovery of DNA – the secret of life – to the world at large on 28 February 1953, (the 'world at large' being other regulars at a local pub). But it is well enough documented that it happened in *The Eagle* in Cambridge at lunchtime on the above date.

The final breakthrough came that morning, possibly a couple of hours before, but for our purposes it was still on the day of a full moon.

We can see the full moon symbolism everywhere – the flowering of illumination; the emotional outburst in a watery tavern; even the connection with ancestry that DNA illustrates.

On the composite chart of Watson and Crick, Uranus, the planet of breakthrough and revelation is within a two-degree orb of this full moon.[17] And this full moon at 9°54' Virgo/Pisces is at that special place at the antiscion of the Sun's exaltation at 19+ Aries, that appears to have global implications. [As mentioned in 'Winds of Change'].

Although the full moon story of the discovery of DNA will forever have Watson and Crick's names entwined in its double helix, the saga had its own 'dark lady' in the person of Rosalind Franklin, now officially acknowledged as an important early collaborator in the scientific quest. By all accounts others found her difficult to liaise with and she chose to leave and work at another location on a different project some time before the final leap was made. It was her immediate colleague and superior at King's College London, Maurice Wilkins, who called her 'the dark lady' in a letter to Crick, but it was Watson in his colourful bestseller *The Double Helix* published in 1968, fifteen years after the discovery and ten years after Rosalind Franklin had died, who described her as a bad-tempered person who hoarded her data.[18] Rosalind Franklin's birth chart (25 July 1920, London, time not known) has Saturn at 9° Virgo, conjunct the full moon of this discovery chart. The dark lady, who died of cancer at the age of 38 and never lived to share the Nobel Prize with Watson, Crick and Wilkins, hovers as an awkward shadow under the brilliance of this full moon. It was assumed by many in later decades

---

17. James Watson: 6 April 1928, 01.23 CST, Chicago, IL, USA. Francis Crick: 8 June 1916, Northampton, England. Time not known.
18. *Rosalind Franklin. The Dark Lady of DNA* by Brenda Maddox, HarperCollins, London, 2003. James Watson subsequently insisted that Franklin's name should be credited along with their own

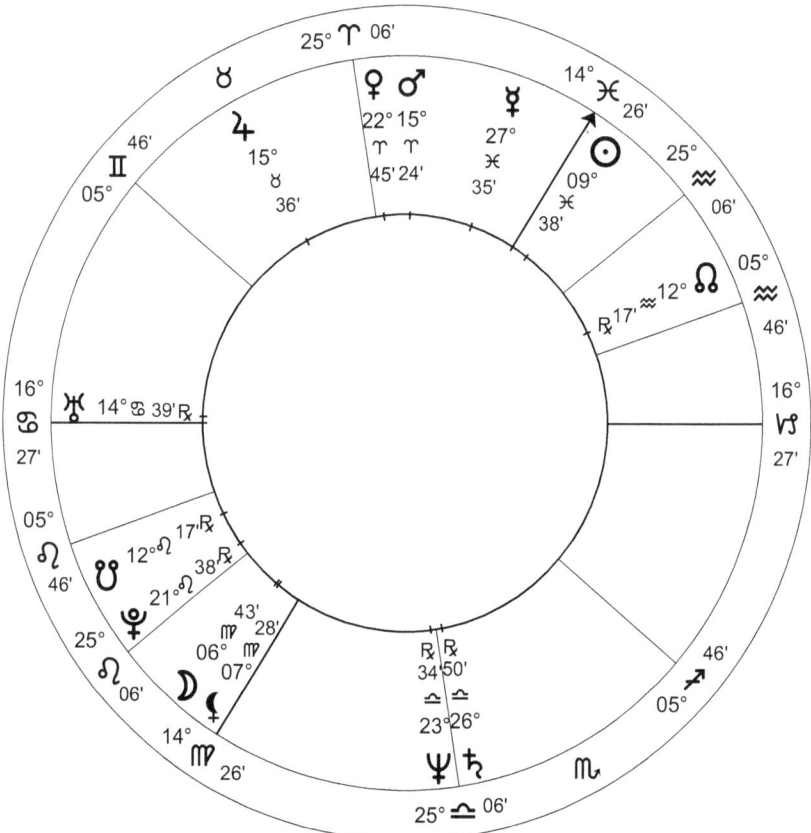

The Discovery of DNA: 28 February 1953, Cambridge, England.
Approx 12.30pm.

that male-dominated Science had side-lined her contribution on purpose simply because she was a woman. That was not the case but note how the Black Moon Lilith is conjunct the White Moon on this chart. The dark lady is unquestionably standing there on this day of illumination. This is a full moon coinciding with the moon's monthly apogee – a strong dark lady indeed. [See more in Black Moon section].

The Eurovision Song Contest
A Sagittarian full moon was rising over a Swiss lake at the first ever Eurovision Song Contest held at Lugano, Switzerland on 24 May 1956. The closeness to a large stretch of water was appropriate, as was the winner of the first ever contest being a thirty-year-old woman. Lys Assia, who sung

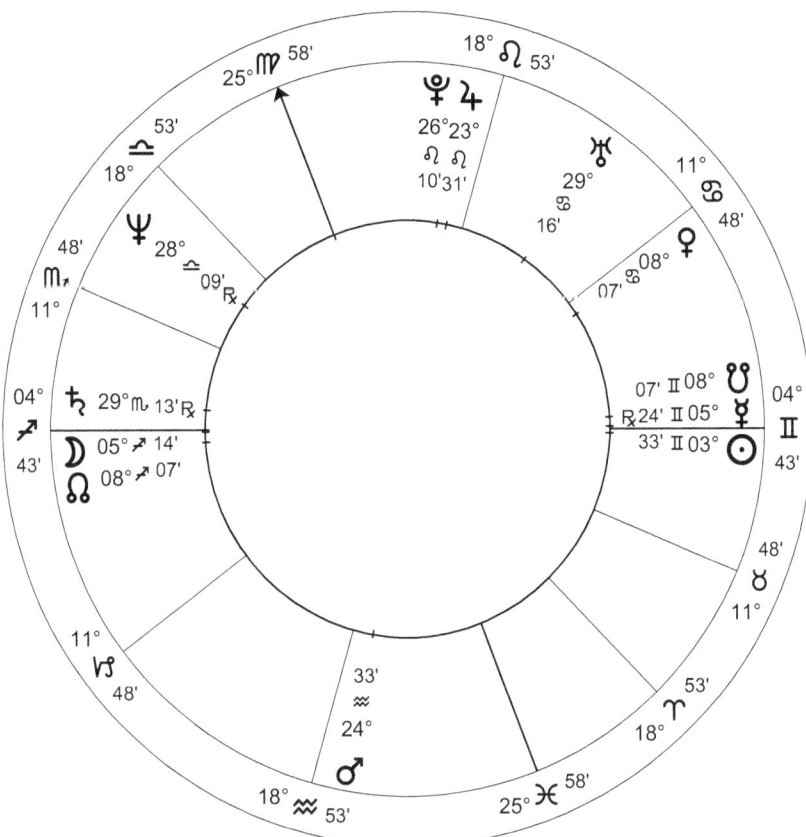

The Eurovision Song Contest: 24 May 1956, 8pm CET,
Lugano, Switzerland.

a nostalgic song of sadness and regret, was the age of one progressed moon cycle. It was a lunar eclipse too,[19] so extra powerful.

Jupiter as both moon and chart ruler described the multi-national, multi-lingual event that would gradually expand from a handful of mainland European countries into one of the largest annual involvements in the history of television. Its expansiveness foreshadowed a growth truly Jupiterian in scale, (chart-ruling Jupiter conjunct Pluto in Leo in the ninth), for in later years there were simply too many countries and contestants to fit into a show that now demanded an entire evening's viewing. Lunar patriotism and illogical voting systems could have marred this European cultural

---

19. Partial not Total.

phenomenon but Jupiter usually managed to keep everyone smiling.

It all grew beyond the wildest dreams of those who instigated the concept of 'Eurovision'; originally conceived in the 1950s as a serious broad-ranging attempt to unite the nations of Europe through the new medium of television. In the public mind these grand schemes may not have counted for much but its least likely element, the cheery old song contest, goes marching on.

Probably the most celebrated and influential of all the winners of the Song Contest over the years were Abba, who won on 6 April 1974. This was one of the very few dates on which a full moon again fell on the night of the competition, in this case near the climax of the show itself, just after 10pm local time.[20]

There has occasionally been a full moon within twenty-four hours of the contest and some of these did have their moments too. 1985 had a full moon that day (4th May) – again a lunar eclipse, this time in Scorpio – and the winning act came from Norway, traditionally a Scorpio country. It was also the year in which the dress of the Swedish hostess fell off during the course of the show. As she was appropriately covered in the right places underneath, the audience – after its initial gasp – took this to be just a contrived piece of theatre. But in fact it was a genuine accident. Her skirt had caught on a nail as she returned to the stage for the voting. Or so we were led to believe. When the full moon fog cleared and clarity of thought returned the Eurovision ruling body decreed that unrehearsed 'accidents' would no longer be tolerated. This did not prevent the 2003 Turkish winner getting her heel caught in a grating on the stage at the climax of the 2004 show in Turkey as she was trying to hand over the trophy to the Ukraine.

If anyone wondered why a grating would be built on the side of a television stage, or in fact why anything made a great deal of sense in this moon-flavoured competition at any time, they would doubtless shrug and mumble 'Eurovision' – a useful by-word for the full moon.

Benny Hill

On the night following the birth of the television comedian Benny Hill (1924-1992) there was a full moon at zero degrees Leo/Aquarius, squaring Saturn. At his birth the Moon was in Cancer. It was a strong Moon whichever way you look at it.

---

20. 6 April 1974, 10.02pm BST, Brighton, England. There was also a bomb hoax that night.

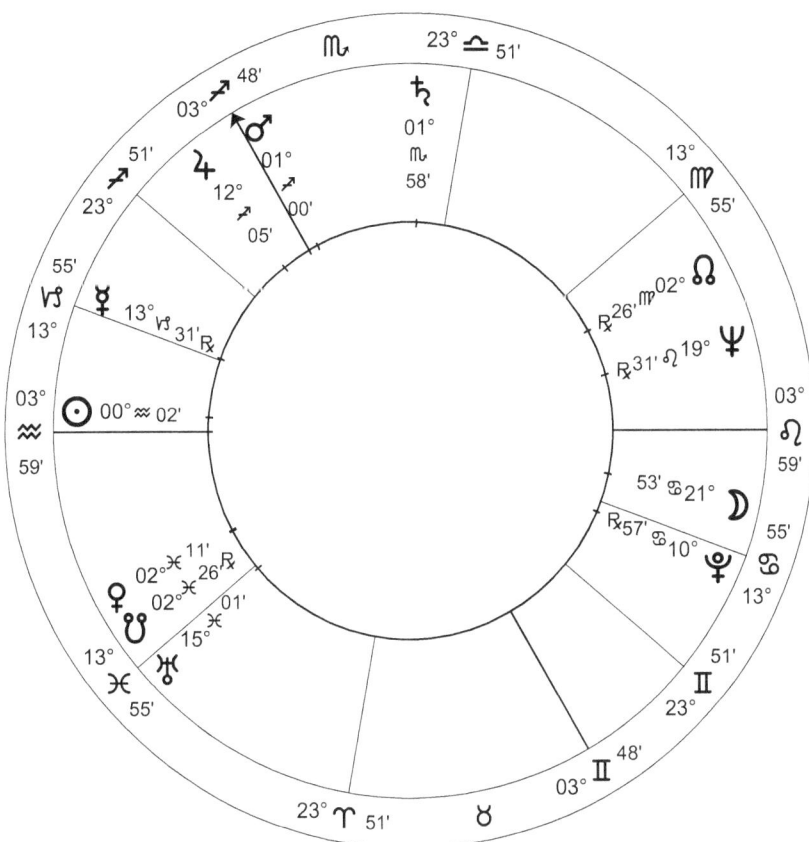

Benny Hill: 21 January 1924, 8.15am, Southampton, England.[21]

This round-faced man who loved women left conflicting legacies in the public mind.

Praised by the likes of Bob Hope, Jack Lemmon, Mickey Rooney etc. as a master comic, he was scorned from different quarters as an emotionally infantile ageing host, leering after women in states of undress. It was the uncomfortably out-dated themes of the latter that led *The Benny Hill Show* to be finally dropped from British Television in 1989 after running in various forms through four decades.

In his early television shows, Benny Hill – then much younger – often impersonated women quite realistically; copying their mannerisms to the hilarious delight of the women in the audience. This was not cruel

---

21. Source: Frank C. Clifford, *British Entertainers*, Flare, London, 1997

or acutely sexist, it was just another of his many impersonations. He had largely pioneered the television format of quick one-joke sketches involving many different characters, something new at that time that neither stage nor cinema could match. Benny had never been much good on stage but this newfangled, futuristic, electrical gadget called Television was right up his Aquarian street. Beside the full moon madness Benny Hill, with the Sun and ascendant in Aquarius, was naturally eccentric. In later years with a worldwide audience and a multi-million income (Jupiter in 10th), he still lived alone in a modest rented flat, went shopping with carrier bags from his local supermarket, and lived with a minimum of material possessions. (Sun and Moon in 'selfless' houses, sixth and twelfth).

Much has been speculated on his enigmatic relationship to women and we have only his birth chart to guide us into a great sensitivity in this area. With his Moon in Cancer and his Venus exactly conjunct the South Node in Pisces, it is possible to read a karmic reverence towards women as the maternal, fairer sex. One of his many jobs on the road to fame was as a milkman (moon-man), famously parodied in a big hit record *Ernie, the Fastest Milkman in the West*. Throughout his life Benny Hill never married and the few women who did know him described him as shy and kind. But his desire for the reputation as a saucy joker surrounded by pretty girls may stem from a Jupiter-flavoured Mars on the midheaven squaring Venus on the nodal axis. Aquarius can often shock others quite unintentionally, but this chart shows a private sensitive man – he hardly ever gave interviews or made live stage appearances – who was at the same time friendly with ordinary people (Aquarius) and loved children (Cancer). He adored women too but his avoidance of live audiences stifled his ability to gauge the public's changing reaction to risqué jokes over the decades. He had crossed a borderline from bawdiness to voyeurism and an increasing female backlash towards his shows gained momentum.

When the sexual powder keg exploded and his television contract was rescinded he was devastated. He had believed he was loved by the public and he died within three years of this massive blow to his perceived popularity. In fact his life ended in an armchair in front of a television. It is assumed he died of a heart attack in his flat on 18th April 1992, the day after a full moon, for his body was not discovered until two days later. Like a favourite mistress shedding tears, the television was still flicking images on to his lifeless body as a neighbour and the police broke down the front door.

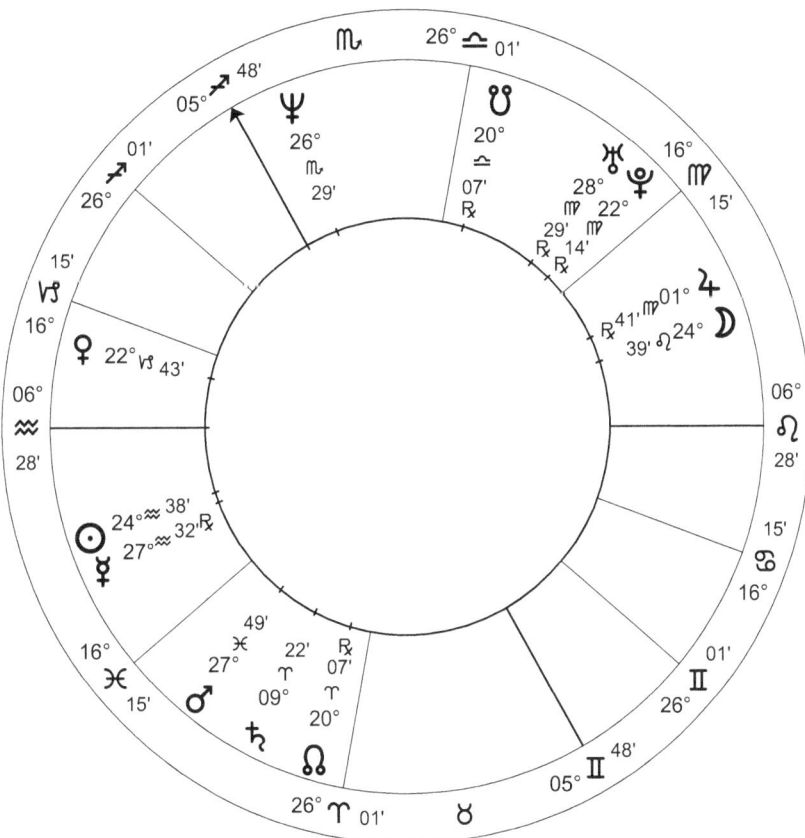

Lara Croft: 14 February 1968, 06.43 GMT, Wimbledon, London. 51N25

Lara Croft

It was only a computer game and she was just a computer graphic. Yet the tomb-raiding adventurer Lara Croft once had clubs and pin-up calendars in her honour, pop tours named after her and products repackaged to reflect her glory. She was one of the biggest female stars of the 1990s/early 2000s, inspiring big-screen movie-films actually featuring 'real' people; but who was she?

Lara was first created by the artist Toby Gard in 1994 in the form of a pencil sketch, and later marketed as a computer-animated character. According to the vital statistics available on her website (laracroft.com), the bosomy daughter of Lord Henshingly Croft was born in Wimbledon on 14th February 1968. No time of day is given but the ephemeris shows that this was the day of a full moon at 24 degrees Leo/Aquarius, and it is the time of the full moon at 06.43 GMT that has been used as the basis of our chart.

Some individuals loved Lara with an obsessive passion. There is certainly a touch of the Leo full moon about all this. Many became addicted to manipulating her through the dangers she faced in her fantastic tomb-raiding quests. (Neptune in Scorpio T.squares her full moon from the ninth house).

Although the adventures of Lara Croft were supposed to be based in our world, particularly Egypt through which her astrocartography lines run, she is clearly not in quite the same universe that most of us know. The whole genre of early computer games involved the player directing him or herself in the form of a fantasy figure through various ordeals often through different levels or 'worlds' ultimately to retrieve or rescue an object or prize. It sounds like a lunar or shamanic journey heightened here probably by Mars in Pisces aspecting many planets on the chart. Lara Croft is a kind of shaman herself – a mediumistic full moon person who allowed others to inhabit her mesh of 38,000 high texture polygon morphing cells and then to travel onwards via her body. The risk, as Charles Carter might have put it, was that Mars afflicting the mental rulers (in this case Mars opposite Uranus and inconjunct Mercury) may bring a subtext of obsession or insanity, and *"is especially dangerous if in a watery sign"*.[22] No wonder that Lara was blamed for causing marital break-ups and job losses for those obsessed with entering her world and staying there for days and nights on end.

The Leo full moon was being transited by Uranus when the film *Tomb Raider* starring Angelina Jolie[23] was on general cinema release in 2001. Neptune was transiting Lara's ascendant at this time too, so it is interesting to observe how such energies work on virtual-type people who are only computer graphics: it made her real instead of unreal. Perhaps everything works in reverse in her world and Uranus transiting her Sun and opposing her Moon turned Lara Croft into a predictable, establishment figure.

---

22. C.E.O. Carter, *An Encyclopaedia of Psychological Astrology*, TPH, London, 1963. First published 1924.
23. Angelina Jolie: born 4 June 1975, 09.09 PDT, Los Angeles, USA. Her chart makes no obvious connection to Lara's except that both have Moon-Jupiter conjunctions

The Old Moon

> 'All things that pass
> Are woman's looking-glass;
> They show her how her bloom must fade,
> And she herself be laid
> With withered roses in the shade;
> With withered roses and the fallen peach,
> Unlovely, out of reach
> Of summer joy that was'
>   [Christina Rossetti, 1830-1894.
>   From *Passing and Glassing*]

The Old Moon

'Emptiness is a great feminine secret' (Carl Jung).

The old moon is poetically the witch, the hag, the wise woman, the fairy godmother.

The opposite of the word 'full' is 'empty', and although we don't speak in English of an empty moon as we do a 'full' moon, it is still highly descriptive of the moon's waning phase, especially the last days of the lunar month when the waning crescent thins into obscurity. Rightly this is a time of emptying, of divesting and relinquishing and making space for the energies of the new moon soon to come. It may be the dark before the dawn, it may be an eggshell broken and hollow, but the phase has its necessary place for withdrawal and ending.

The darker the moon the less value there is in starting something new. The advantage of this time is the ability to consciously be rid of unwanted attachments of misfortune or illness. It is the time for clearing the ground, lying fallow, removing debris and symbolically shedding skin.

The old moon is the reaping end of the cycle, the season of sorting the wheat from the chaff and sifting through the wisdom gained. It is the seeding time, when the flower and fruit of the earlier phases has run to seed. The popular imagination doesn't much like this period, hence the adoption of the word 'seedy' to describe something shabby and old and past its prime. But life goes on and the new cycle will be born on the back of the last. We can carry through the best of what has been harvested but should prune, simplify and get rid of burdens with the old moon.

Do you have a decorative box somewhere in your home that remains empty because you can't think what to put in it? This is your ideal Moon box.

The moon is a natural container and its period of emptiness has established its before-the-dawn status. Eighteen centuries ago Plotinus observed that to attract an aspect of nature one should fashion a container designed to receive it. He was referring to ancient shrines, but nature will always be attracted to its own. Now you can deliberately deposit in the box an idea that you want to grow. No moon remains in the same phase forever; it needs seeds to nurture in the dark. Add silver coins to your box, add a written wish on a card decorated with moons – add whatever you like, and watch what happens in the coming lunar months.

The Cinnamon Tree in the Moon

There is a Chinese legend that tells of a Cinnamon tree that grows on the moon whose roots and branches relate to all the peoples of the Earth. This vast family tree is pruned once a year and on the fifteenth day of the eighth month some of its branches fall to Earth. The pruning aspect likens it to an 'old moon' ritual.

If you are lucky enough to find a fallen branch on this special night you should form or incorporate it into some kind of empty container – a moneybox, a treasure chest, a wardrobe… and deposit a sample of your need inside, for over the following year, the story promises, it will bring an abundance of what it is designed to hold.

What is not clear in the telling of this fable is whether the fifteenth day of the eighth month means the 15th of August, which would roughly coincide with the annual Perseids meteor shower, or whether we are speaking of lunar months. The fifteenth day of the eighth month – or of any lunar month – would be a full moon and the 'eighth' month would presumably be the eighth lunar month counting from the first new moon in Aries. This makes the eighth month a Scorpio new moon starting some time in October/November, and the fifteenth day or full moon (in Taurus/Scorpio) taking place annually in November. It's an odd concurrence that these times of the year coincide with the other two famed meteor showers – the Orionids in mid to late October and the Leonids in November.

At their most dramatic, meteor showers can appear like many silver sticks or branches raining down from the heavens.

Some OLD MOON charts

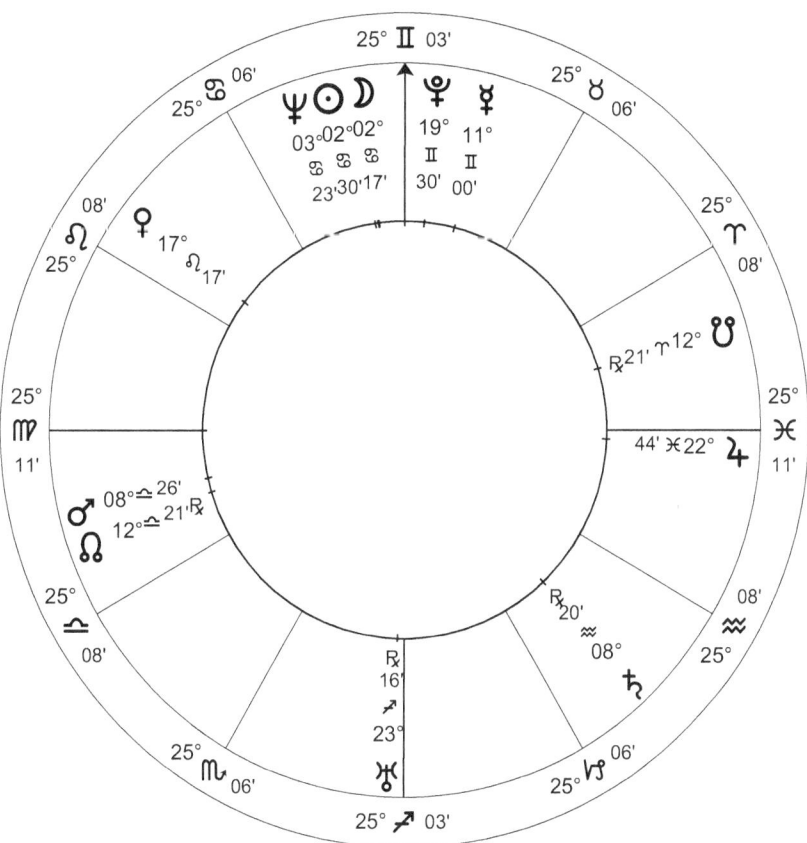

George Orwell: 25 June 1903, 05.50am GMT, Motihari, India].[24]

George Orwell

George Orwell's vocation and success as a writer can be easily seen in this version of his chart with a Gemini midheaven and the Sun and Moon conjunct it in the tenth (and Mercury as chart-ruler and part of a Grand Air Trine). They say the darkest hour is before the dawn and the man who wrote his nightmare visions in a novel called *1984* had the darkest moon possible. It was in the last thirty minutes (of time) before a new moon and conjunct the fantasy planet Neptune.

---

24. Source: AA Database gives their source for the birth time 'Unknown'. Lois Rodden in *The American Book of Charts* gives 11.30am LMT, which produces the same ascendant as the AA's GMT time, but she labels it 'DD' – unsubstantiated/speculative.

The extreme introversion of the dark imaginary world of *1984* where hidden cameras watched your every movement, even from trees in forests, and all free thought was denied by the State, is actually quite an illogical idea. At the time the book was written in the late 1940s such preposterous technology would require one person to be constantly watching one other person for twenty-four hours non-stop; a little difficult for a society to function if everyone was watching everyone else surely? It was a futuristic science fiction allegory of course but the point to be made is that it's a totally lunar concept in its most fearful and inward-looking form. A society gone insane. We note from Orwell's chart that the moon is the sun-ruler and in its own sign. It is a powerfully strong moon although it is dark and old.

The Moon Under Water
In 1946 George Orwell wrote a short essay, later to become famous, on his ideal of the perfect pub, which he called 'The Moon Under Water'. This fictitious public house had motherly barmaids and creamy draught stout, warm open fires in winter and a secluded garden out the back in summer. It was no different to a hundred similar inns and taverns up and down the country but it captured the essence of all of them and exuded a sense of security with which most of his countrymen and women identified.

Pubs are cosy little moonlike places whose primary purpose is to serve liquid. Traditional pubs have doorways like normal houses, which enter into room-size bars. They are quite rightly public 'houses' or homes, very different to a high street saloon with its shoplike frontage and neon lights. Foreign visitors are usually surprised to find that the orderly rules of the outside world do not apply once you step across the threshold of a British pub.

The most obvious cultural reversal is that unlike every other public venue, no one queues to be served in a public house. British society tends to form orderly queues at every available opportunity, reflecting perhaps the Capricorn Sun on the United Kingdom chart, but these self-imposed regulations do not apply once inside a crowded gossipy pub. If you stand politely waiting for your turn to be served the chances are you will never get a drink all night. It may be due to the full moon in Cancer on the UK chart. In a homely atmosphere one has no need to bow to convention; inhibitions can be shed and emotions flow freely.

Orwell's essay was an article written for *The Evening Standard*, a London newspaper, and was first published on the evening of 9 February 1946. If we assume a time of around 6pm for its appearance on that date we find that the Moon was exalted in Taurus and riding high in the chart. This however was a first quarter moon rather than a waning one.

Punch
Punchinello is a very moon-faced fellow. His nose and chin curve round side-faced to resemble a crescent. He has a humped or hunch back. His loony antics are designed to make us laugh even though they have an undeniably darker side.

[Punch (after a Victorian doorstop ornament)]

Punch was the name of one of the longest-running humorous magazines in the English language, published weekly from 1841-1992. It was originally named Punch after the drink with a kick, but the double meaning of comic Mr Punch sealed the inspiration and the cover of the first issue showed a crowd gathered around a Punch and Judy show. The last ever cover showed Punch and Judy as senior citizens hobbling off into the sunset. Perhaps the abiding image for an old moon.

The magazine had always been a magnet for the ablest of comedy writers and caricaturists, and it is credited with inventing the format of the punch-line cartoon that became standard everywhere. The word 'cartoon' itself, which before the mid-nineteenth century had simply meant an artist's sketch design, took on the new implication of a satirical humorous drawing. The first issue of *Punch* magazine appeared on Saturday 17 July 1841 selling 10,000 copies. Its highest circulation peaked in the 1940s, but numbers declined as with all newspapers in the television age and after a last revival in 1996 it finally ceased publication in 2002.

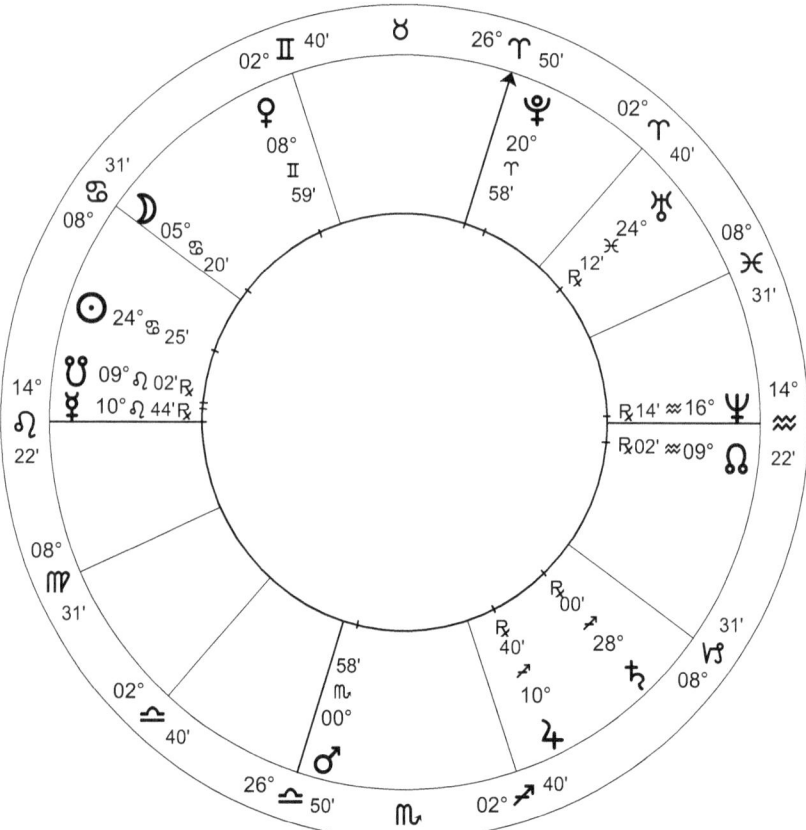

Punch magazine: 17 July 1841, 06.00am, London, England

The chart is set arbitrarily for 6am, a rough guess at the time the first magazine may have been available to buy from a news-stand, but we shall give no weight to the angles or houses or the actual degree of the Moon which at this daybreak time would have been trining Mars and opposite Saturn. At whatever time we take on this day with its fortuitously anti-establishment Sun trine Uranus, it is an old moon waning in Cancer, anarchically out-of-bounds.

The moon is old but wilfully strong. Named after a drink and a long-established children's entertainer – Mr Punch belonged to a tradition that stretched back to 17th century Italy – we can see the Cancer old moon themes. Eating and drinking were an important part of the original *Punch* publishers' lifestyle and their staff meetings were always held in a pub. The 'Punch Table', soon relocated to its own offices, became not only an excuse

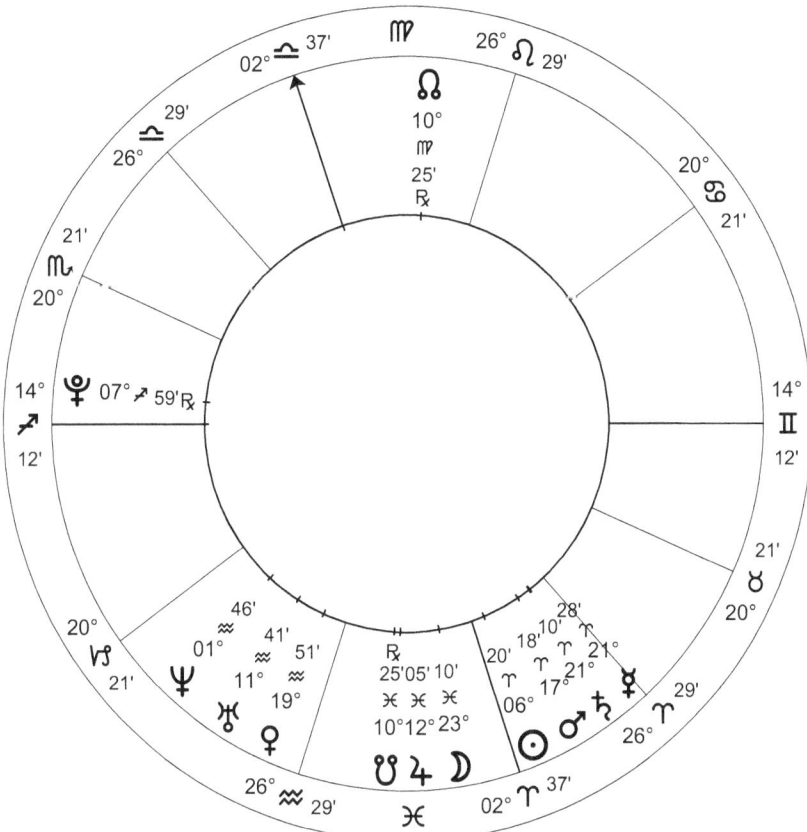

Viagra: 27 March 1998, 00.00hrs EST, Washington DC, USA

for a regular noisy banquet but a well-loved custom that acquired its own prestige and was continued by editors through the decades.

A great deal of nostalgia is held towards *Punch*. Its unrivalled reserve of cartoons remains a historical resource spanning a century and a half of changing fashion, politics and attitudes.

Viagra

You don't need Viagra when you're young and virile; it's an old man's solution and not surprisingly we find the moon in its dying hours on the date that Viagra was approved in the USA for public use.[25] Washington DC is taken as the symbolic place, zero hours as the time.

---

25. Approved by the US Food and Drug Administration for public use from 27 March 1998. Source: *80 Days that Changed the World*, Time Inc, New York, 2003. Washington DC taken as symbolic place; zero hours as time.

Note Mars in Aries is conjunct Saturn, and Jupiter (expansion) is on the South Node.

The Moon Was Pluto – or Pluto Was The Moon

Astrologically speaking, the moon in its dark phase once ruled much that later came under Pluto's wing.

Frogs and toads for instance, those ugly watery magical creatures whose life cycle illustrates transformation, are common to both Pluto and the dying/renewing moon. Traditional fairy stories are full of death and rebirth – and frogs and toads – and these were stories once told to children to prepare them for life's changes. Children's stories are moon fables, bringing them comfort and security with the promise of a 'happy-ever-after' despite the dangerous taboos and disturbing things that are encountered on that road to maturity.

Swamps, marshes, dripping grottoes and dank sunless caverns descending deep into the fissures of the earth, can all be seen as belonging to either the Moon or Pluto – rulers of embryos and abortion. Ancient cultures unearthed through excavation, sunken galleons raised from the sea and fabulous treasure mines coming to light follow a similar rulership arrangement. Then there is the lurking darkness of occultism and black magic – once ruled by the dark moon, later by Pluto. And although Pluto in mythology is king of the underworld, Pluto in astrology has often been given a feminine side, ruling grandmothers for instance.

Common to both planets are evolution; gestation and birth (Luna), transformation and rebirth (Pluto).

As a celestial orb the moon is not greatly different in size to Pluto. The diameter of Luna is 2160 miles, or 3476 kilometres; the diameter of Pluto is 1430 miles, or 2302 kilometres. One of Pluto's moons was renamed 'Nix' in 2006 after the goddess of night (Nyx). On the chart of Pluto's discovery as a planet in 1930 the Moon was in its last quarter, in fall in Scorpio and conjunct its south node. That's a fairly dark moon, appropriately. Seventy-six years later when Pluto was demoted from full planetary status to 'dwarf planet' by the International Astronomical Union in 2006 the Moon was new, less than a day old. The Moon was free of her previous identification with Pluto – is that what it was saying? Pluto on this demotion chart was placed at 24° Sagittarius, exactly the IC of its discovery chart.[26]

---

26. Pluto's discovery: 21 January 1930, 10pm MST, Flagstaff, Arizona, USA. Note: There are alternatives to this chart. The one referenced here represents the

The Witch

She dresses as dark as the midnight sky in billowing cloak and pointed hat. She's a hump-backed hag, riding a broomstick with a black cat behind her. They're flying across the face of the moon. She's old and ugly; warts on her nose and bristles on her chin...

This harsh mimicry of the old moon in its most waning phase first caught the public imagination in this form in the seventeenth century, although witches have existed since the earliest of times.

The old moon is the wisdom of the dark; uncomfortable to most. It's the wisdom of age; denied by most. And it's *lunar* wisdom with everything that that implies. The old crone dressed in black may be a parody of the moon at its darkest, but it is based on a figure once respected and feared. Respected for the uncanny knowledge she has acquired; feared for it being unknowable and possibly dangerous.

Coming back to popular misconceptions: Why do we see a witch this way? The black clothes are plainly aligned to the darkest phase of the moon; the flying aspect reminiscent of the Valkyries – but the pointed hat, the warts?

Magicians wear pointed hats in popular thought, as do dunces in the corner of Victorian schoolrooms. Wise men, fools and witches all wear conical hats therefore, a remnant no doubt of the belief in funnelling and amplifying a spiralling energy of thought into the brain. The witch's hat would be receiving the wisdom of the moon in this fashion. This type of hat appears historically as the cap of the haruspex, an augur or diviner, on pre-Christian Etruscan bronzes. Ancient coins from the city of Luna in Etruria, modern-day Italy, show a head wearing a conical hat that may represent Artemis/Diana herself.

Warts. By superstition warts were tied to the moon's phases. Rustic folklore recommended buying your warts with silver – pretending to buy them off your body or getting someone else to, so they didn't belong to you any more – or touching them with something that could be buried at the dark of the moon. As the relevant substance decayed in the earth so did the warts wither and disappear. And because they seemingly came and went like magic, warts were associated with witches, as were all kinds

---

discovery from comparison of photographs. Source: David McCann, 'The Birth of the Outer Planets', *The Traditional Astrologer*, Nottingham, 2000.

Pluto's demotion: 24 August 2006, 11.38am CEST, Prague, Czechoslovakia. Source: Contemporary news reports.

of moles or skin blemishes – 'the Devil's marks' according to witchfinders. On a fictional level the association persists into the present day with Harry Potter's school of witchcraft and wizardry called 'Hogwarts'.

The black cat as the witch's familiar still holds a high place in the Top Ten of popular superstitions. To see a black cat or to have one cross your path is either thought lucky or unlucky depending on the geographical area of your birth; the local native belief. Whether portending good or bad fortune, there is an ominous feeling about a jet black creature staring into your soul with its see-in-the-dark eyes. [More on cats in the Black Moon section]. There is an assumption that cats and witches share a common understanding, perhaps that they can change into each other. Such hysteria mixed with religious fanaticism enflamed the historical witch hunts when thousands of women and cats were put to death, simply for being women and cats. Oddly perhaps the warped minds that dreamed up the rules for determining whether a person was a witch in the dreaded medieval handbook *The Hammer of Witchcraft* did not go harder on the victim if the accused was known to practice astrology. The stars were believed to be in the realm of the angels and their influence could not be warped by evil spirits.

The witch's broomstick made of twigs and sticks was a straightforward household item once found in every stone or wood-floored home. This in itself gives it a moon connection. Prudence Jones in *Northern Myths of the Constellations*[27] mentions an old mural depicting the goddess Frigga riding what could be either a broomstick or a distaff. Both are objects associated with housework. The distaff – a stick used in spinning – has extremely ancient affiliations. Frigga was a fertility goddess and the image of a woman with a phallic-shaped distaff or broomstick under her skirts suggests procreation, as in the custom for newlywed couples to 'jump the broomstick' or jump over a candlestick to encourage a fruitful union.

Modern day witches honour the moon in all its phases as a symbol of the Divine Goddess. The practice of Drawing Down the Moon, usually on a full moon, is a ritual to align the earth with the changing cycles of existence and draw down to the earth the positive aspects of fertility and feminine power. The ancient rite is said to have its roots in Thessaly in Ancient Greece, where witches were believed to actually bring down the moon from the sky.

---

27. Prudence Jones, *Northern Myths of the Constellations*, privately printed booklet, 1991

Some witches of history, fiction and legend

One of the Thessalian moon priestesses was called Aganice, quite probably a figure of historical reality who was famed also as an astrologer. Sceptics have naturally claimed that her ability to draw down the moon was through a knowledge of the timing of eclipses and nights when the moon would disappear so that her more credulous observers could be taken in.

The Witch of Endor

One of the most well-known biblical witches is the witch of Endor, mentioned in the first Book of Samuel in the Old Testament. The story opens with the people of Israel at war with the Philistines and Saul, the Israelite leader, nervous about the outcome. Saul had been getting no helpful answers from his dreams or prayers and had recently "put away those that had familiar spirits, and the wizards, out of the land". But now he found himself in need of guidance.

> 'Then said Saul unto his servants, Seek me a woman that hath a familiar spirit, that I may go to her, and inquire of her. And his servants said to him, Behold there is a woman that hath a familiar spirit at Endor.
>
> And Saul disguised himself, and …came to the woman by night: and he said, I pray thee, divine unto me by the familiar spirit, and bring me him up, whom I shall name unto thee.
>
> And the woman said unto him, Behold, thou knowest what Saul hath done, how he hath cut off those that have familiar spirits, and the wizards, out of the land: wherefore then layest thou a snare for my life, to cause me to die?
>
> And Saul sware to her by the Lord, saying, As the Lord livith, there shall be no punishment happen to thee for this thing.
>
> Then said the woman, Whom shall I bring up unto thee? And he said, Bring me up Samuel.'

The story continues with the spirit of Samuel admonishing Saul for calling him up from unearthly realms and 'disquieting' him. He then tells Saul that the Lord is not talking to him because he (Saul) refused the holy order to carry out the Lord's wrath upon Amalek and therefore Israel will lose the battle with the Philistines and Saul and his sons will die on the morrow. Saul is struck with fear and already weak from lack of food, he falls to the ground in a faint.

> 'And the woman (the witch) came unto Saul, and saw that he was sore troubled, and said unto him, Behold thine handmaiden hath obeyed thy

voice, and I have put my life in my hand… Now therefore, I pray thee… let me set a morsel of bread before thee: and eat, that thou mayest have strength, when thou goest on thy way.

But he refused, and said, I will not eat. But his servants together with the woman, compelled him: and he hearkened unto their voice. So he arose from the earth and sat upon the bed.

And the woman had a fat calf in the house: and she hastened, and killed it, and took flour, and kneaded it, and did bake unleavened bread thereof.

And she brought it before Saul, and before his servants: and they did eat. Then they rose up, and went away that night.'

And that's the last we hear of this kindly witch. The next chapters go on to describe the battles and the pillaging and how Saul and his sons died as prophesied.

Mother Shipton

'A weird and wicked-looking hag,
With cheeks as flabby as a bag…' [28]

In the north of England in the reigns of King Henry VII and VIII, there lived a Yorkshire prophetess called Mother Shipton.

Rumoured to have been born in a violent storm in a cave in Knaresborough (now a tourist attraction with a petrifying well – literally petrifying, it turns objects to stone), her visions supposedly foretold many coming events in world history.

There's no birth chart for Mother Shipton, no reliable birth date at all, although there was once claimed to be a gravestone in Clifton that marked her burial place. (With no date on it). One old source, *Mother Shipton and Nixon's Prophecies*, published in 1797 and supposedly compiled from older editions, states she was born in July 1488, baptised under the name of Ursula Sonthiel and married to one Toby Shipton at her age of twenty-one. But the same book says she died in 1651 at the age of about seventy. This surely means 1551, if there is any truth in it at all.[29]

---

28. From a satirical nineteenth century poem quoted in *Mother Shipton Witch and Prophetess* by Arnold Kellett, George Mann Books, Maidstone, 2002.
29. See *Mother Shipton Investigated* by William H. Harrison, published in London in 1881. 1881 is the date the world was due to end according to one of Mother Shipton's alleged predictions and Harrison's book is available now as an internet archive resource on www.archive.org/details/mothershiptonin00hargoog

Her reputation was such however that almost a hundred years after she probably died her name was still a byword for accurate prediction and the astrologer William Lilly reprinted a version of her prophecies in 1645 leading to their even wider recognition. Her visions supposedly foretold the Great Fire of London, the English Civil War, the union of England and Scotland and so on. Other hands added bits in later years fabricating and updating her predictions to encompass further important happenings on a wider worldly stage.

Mother Shipton is important in a socio-historical context because as Arnold Kellett points out in his study published in 2002[30] the earliest artists' impressions of her in the form of woodcuts for the pamphlets of her prophecies became the classic representation of a witch in common thought. These illustrations were almost certainly imaginary; merely artists' impressions of a scary old sibyl. Although Mother Shipton remained a popular figure, the image of this wart-faced old woman in black with a vaguely pointed cap (later to become more like the black high hat still preserved in Welsh national costume) is the enduring caricature with which we opened this section on The Witch. And even more fascinating is that the 17th century artists chose to illustrate her face in what we would now call Mr Punch style with a hooked nose meeting an upwardly curved chin. Her face is therefore a crescent moon shape, as is her hunched posture.

Moths are ruled by the moon and the variety *euclidia mi* is still known in English as the 'Mother Shipton Moth' from the pattern of her profile on its wings.[31]

The Wicked Witch of the West

The evil witch image was undoubtedly reinforced in the popular mind in more recent times by the portrayal of the 'Wicked Witch of the West' in the 1939 movie film *The Wizard of Oz*. This witch was dark robed with a pointed hat, a green face, and a broomstick, although in L. Frank Baum's original novel she mainly carried an umbrella, (because she was afraid of water). However, the movie-film Witch of the West was more archetypal in appearance than W.W. Denslow's illustration for the 1900 novel, which

---

30. *Mother Shipton Witch and Prophetess* by Arnold Kellett, George Mann Books, Maidstone, 2002.
31. This can be as elusive as trying to see a 'hare' in the moon, but appropriate nevertheless.

showed an ugly shortish one-eyed woman with sparse hair plaited into three pigtails.

It's a sign of the times that the character of the Wicked Witch of the West has been reappraised in the 1995 novel by Gregory Maguire and the subsequent hit musical *Wicked* as a truth-seeker in a corrupt land. She may be green and seen as different but she is fair-minded and clever. Witches, like the old moon it seems, have been misunderstood for too long and we are beginning to think again.

The Salem Witch Trials
It's almost certain that there were no knowledgeably practising witches in Salem and the surrounding area of Massachusetts in the late seventeenth century, yet a religious paranoia swept through the inhabitants causing nineteen innocent people, mainly women, to be hung for witchcraft, one man to be tortured to death and at least five others to die in prison.[32]

Arthur Miller's play *The Crucible* (1952), based on the records of the original trials, showed the protagonists as what they most probably were, ordinary decent folk whose strict puritan upbringing mixed with the normal human gullibilities and jealousies had unexpectedly fatal consequences. In retrospect we see a whole community suddenly moonstruck, as it was once called, with those in authority duped by adolescent girls accusing each other of ungodly practices and the whole incident rife with emotional madness. The play's title, the 'crucible', referring figuratively to a heated melting pot of a trial, is literally a dish-shaped moon-ruled container. A cauldron. Twenty years after the trials the government gave compensation to the survivors who had been imprisoned and to the families who had been robbed of their loved ones, and the original jury prayed for forgiveness. It's no accident that Arthur Miller wrote his play at the time of the McCarthy communist 'witch hunts' in modern 1950s' America.

Arthur Miller's natal Moon was in Aquarius, waxing gibbous. His wife of five years, Marilyn Monroe reflected this with her own Aquarian Moon *waning* gibbous. Their composite chart has an almost exact full moon (Sun Leo, Moon Aquarius).[33]

---

32. The birth chart of the acceptance of Massachusetts into the Union, 6 February 1788, has the moon in its darkest phase. Carolyn R. Dobson in *Horoscopes of the US States and Cities*, AFA, Tempe, 1975, sets the chart for local noon.
33. Arthur Miller: 17 October 1915, 05.12am, New York, NY. Source: Rodden. C rating. Marilyn Monroe: 1 June 1926, 09.30am, Los Angeles, CA. Source: Birth certificate.

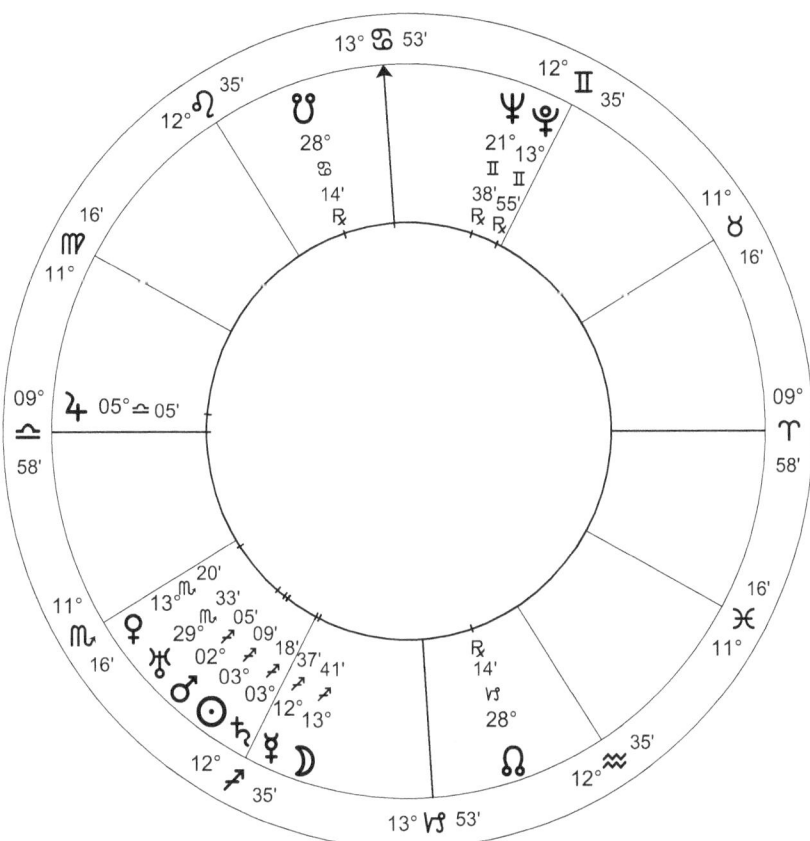

Helen Duncan: 25 November 1897, 3am, Callendar, Scotland

Helen Duncan

Helen Duncan was the last so-called witch to be prosecuted and sent to prison in the United Kingdom in a trial held in 1944. Actually she was a spiritualist medium. Many spoke for her good character at the trial but the jury found her guilty of falsely proclaiming to procure spirits and she was imprisoned for nine months. Winston Churchill himself expressed private disgust about the whole affair calling the charge "obsolete tomfoolery".[34]

Helen Duncan was born in Scotland, hence her birth time is accurately recorded on the birth certificate: 3am, 25 November 1897 in Callander.[35]

---

34. A copy of Churchill's memo to the Home Secretary asking for an explanation as to 'why the Witchcraft Act of 1735 was used in a modern court of justice' can be seen on the Helen Duncan website. www.helenduncan.org.uk
35. Birth times are officially recorded in Scotland but not in England.

This new moon chart with Sun and Moon in Sagittarius suggests that Helen was straightforwardly open about her business, perhaps too much so eventually causing the Sun conjunct Saturn to exact a heavy price. Her out-of-bounds Moon rules the midheaven and therefore her unusual reputation and measure of fame as a medium. The Moon is also conjunct Mercury and opposite Pluto (exact); suitable symbolism for receiving messages from the dead.

That a genuine medium should be brought to trial and imprisoned in the middle of the twentieth century had more to do with it being wartime than any quarrels about her psychic abilities. Helen, then situated in the naval town of Portsmouth, was innocently passing on in trance state messages from sailors who had died before such classified information had been officially recorded, or admitted. The authorities became suspicious and she was arrested and held as a spy. This charge was later changed to incorporate something from an ancient Witchcraft Act, but it is now assumed that there were worries she might have picked up and subsequently given out the highly top secret details of the coming D-Day landings and it was in the interest of the realm to keep her safely out of the way for a while.

These outdated Witchcraft and Vagrancy Acts, which could also apply to astrologers, were not finally repealed in the UK until 1989. The last British astrologer to be brought to court under the Vagrancy Act was Alan Leo in 1917. In the United States the celebrated trial of Evangeline Adams in New York in 1914 has become even more legendary as she was said to have defended herself dramatically in court and was subsequently acquitted.

Bedknobs and Broomsticks
Mary Norton's three children's novels about the amateur witch Eglantine Price were also written during the Second World War although the conflict is not overtly mentioned. When Walt Disney later filmed a version of the stories as *Bedknobs and Broomsticks* (1971) it was specifically a tale of three young evacuees from war-torn London staying with a dear old lady in the country who was endeavouring to master a correspondence course on witchcraft and use it to foil and repel the enemy from invading.

Both books and film were light comedy but based on the deeper truth that many occult groups in Britain during World War 2 comprised women actively fighting the darkness of Nazism on an inner plane. Other fictional novels that drew more seriously on this include Dennis Wheatley's *Strange Conflict* (1941) and Katherine Kurtz' *Lammas Night* (1983).

Sabrina and Samantha: Two television witches
Thirty years separated two popular television comedy shows about witches: *Bewitched*, first series began 17 September 1964, 8.30pm, New York, and *Sabrina the Teenage Witch*, first series began 7 April 1996, 7pm, New York.

The formula for both was basically similar. Trying her best to act like a normal human being in the everyday world, the heroine often used her magic inappropriately, resulting in farcical mix-ups all round.

Neither of these light and breezy sitcoms was born on an old moon. *Bewitched* is waxing gibbous, *Sabrina* is waning gibbous, but both have the Moon opposite Venus in charts with a Venus-ruled ascendant. These witches were pleasant and pretty and most of the comedy came from the Moon-Venus contradiction of domestic trouble with female relatives, family mayhem and social inexpediency.

Apart from the disparity in the two characters' ages – Samantha was a married woman, Sabrina was a 16-year-old school girl – probably the main difference between the two shows was the change in attitude towards women, witches or otherwise in the thirty years between the 1960s and the 1990s. From the woman-in-a-man's-world sitcom of Samantha the eccentric Moon-in-Aquarius housewife in *Bewitched*, Sabrina the Moon-in-Sagittarius teenager was a far more independent young lady. Where Samantha the devoted housewife used her spells mainly to try and keep the home and her husband's life running smoothly, Sabrina the adolescent had to restrain herself from doing things that all young teenage girls would love to do, like turning bitchy peers into frogs, or making boys fall in love with her. If the series was trying to preach any moral message to its younger viewers, apart from saying you should be responsible for your actions, there was always the topical message of accepting those who were different

Sabrina lived with two aunts (and a talking cat called Salem) who were the main supporting characters. Just as in John Updike's *The Witches of Eastwick*, and other stories and movies on this theme, it is always three witch women who share a home.

> 'Be of Good Cheer – the sullen Month will die,
> And a young Moon requite us by and by:
> Look how the Old one meagre, bent and wan
> With age and fast, is fainting from the sky!'
> [Omar Khayyám 1048-1123]

THE COMPUTATIONAL MOON

'The sea of Fortune doth not ever floe,
She drawes her favours to the lowest ebb;
Her tyde hath equall tymes to come and go,
Her loome doth weave the fine and coarsest webb…'

    [Robert Southwell 1561-1595.
    From *Tymes goe by Turnes*]

Moon Maths

'Logical' may not be the first word we would attach to the meaning of the moon, and its motion is one of the most mathematically complex to work out in fine detail, but like any heavenly orb its properties can be expressed in straightforward numerical terms.

Primarily it is the moon's cycles, which are the most well-known of any planetary body, that have an ordered regularity that can be measured and relied on for the sustaining rhythms of our lives.

In this section we look at certain aspects of the moon's cycle. We'll look at Eclipses, both solar and lunar. We'll examine the magic numbers of the Moon, the hours of the Moon, the Progressed Moon, the Moon as a final dispositor, some zodiac degrees with moon relevance, and the Lunar Zodiac – the Nakshatras or Mansions of the Moon.

But we will remain ever aware that when we are dealing with numbers and the moon we are working so to speak with a misty mirror where numbers can also assume strange runic shapes. As in higher mathematics the use of numbers becomes increasingly abstract – and meaningful. To translate that into street-level terms: Don't expect to total up a column of figures in your head at full moon and expect them to equal the same as they did at any other time. The following is a popular example of what we might call Moon Maths:

> 'As I was going to St Ives
> I met a man with seven wives.
> Each wife had seven sacks
> Each sack had seven cats
> Each cat had seven kits.
> Kits, cats, sacks, wives
> How many were going to St Ives?'

The solution to this old rhyme, as I'm sure you know, is not 2,802 but One. (It is you who is travelling to St Ives). But there is such a 'crowd' of moon figures here including kits, cats and women coming the other way, we wonder if that is the right answer.

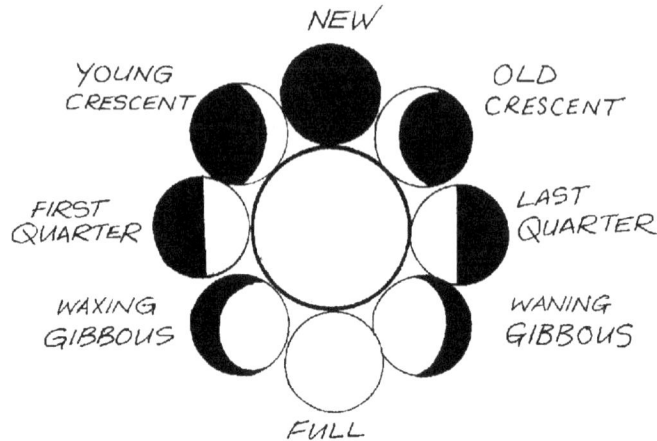

Some say that the answer to the riddle is not One but Zero. That the last two lines of verse should be read together, implying that the question asks how many of the kits, cats, sacks and wives were going to St Ives? The answer in that case is none of them were.

Perhaps the true answer to the whole thing ultimately depends on the way you feel about it.

Moon maths doesn't have to be confusing though…

The Moon's Cycle: The Patterns of Growth and Decay

The moon waxes and wanes, grows and shrinks, from New to Full to Old every month as we have already examined, carrying the tides of life with it. And each new moon begins a lunar cycle often portrayed as an eight-phased journey with each division offering its own special meaning. The work of Dane Rudhyar is generally acknowledged as identifying a specific meaning for each of these eight phases, (see especially Dane Rudhyar's *The Lunation Cycle, A Key to the Understanding of Personality*, 1967).

In a lunar month of 29.5 days each phase of its eight-fold pattern would last just over three and a half days. And the same meanings can be transferred to the progressed moon cycle of 29+ years with each phase lasting three to four years.

Sometimes perhaps it may be counter-productive to think of the changing moon as the hand of a clock clicking into eight distinct and separate phases. The moon is fluid; gradually altering its shape, and the new moon and full moon (and all lunar phases) are only moments in a flowing pattern that has no precise number of divisions.

However it *is* convenient and orderly to see the moon's monthly pattern as an eight-fold diagram and this has become more or less standard in astronomy and astrology.

There are some beautiful surviving old manuscripts of these lunar phases. A Dutch engraving from 1660[1] by Andreas Cellarius labels the Crescent phase *luna falcata* – literally in Latin a scythe-shaped moon, with the Gibbous phase as *luna gibbosa* meaning a hunched or humped moon. Another map in the same collection shows a 12-fold lunar cycle and another with 36 divisions.

The 12-fold diagram is worthy of study as Twelve is a number that incorporates the lunar number Three (3 x 4 = 12). The previous 8-fold design is inevitably all semi-squares, squares and oppositions if you relate the lunar phases to each other, whereas the 12-fold map incorporates trines and sextiles also (see overpage). In fact it incorporates all the major aspects if you measure the angular distances from the new moon to the start of each of the successive phases. You can find yods and all sorts of things. For me there is a fuller feeling of how the particular flavour of each moon is manifesting throughout its monthly course by using the meanings of the aspects as they hark back to the moment of that new moon's beginning. Each phase now has an approximate duration of two and half days, or two and half years if you are working with progression. (See 'The Progressed Moon' later in this section).

Because the 12-fold diagram splits naturally into three, we could say that the active starting-off energy of the new moon in general runs all the way from 'New' to 'Waxing Gibbous', which is ten days roughly. The full moon stage then stretches another ten days from 'Waxing Gibbous' to 'Waning Gibbous', centred around the full moon; and the old moon runs from 'Waning Gibbous' back to 'New'.

Specifically, if we take each of the 12-fold stages in turn, the **New Crescent** is the first appearance in the sky of the thinnest new moon crescent, making a semi-sextile or inconjunct aspect to the moment of the new moon, the start of the cycle. At 30 degrees from the new moon it can be seen to have a 12th harmonic flavour of realistic aspiration, a positive growth towards the ideal. This first phase of the moon in its 12-part cycle is one of the most noticeably moving to the human spirit. Its beautiful physical appearance in the evening sky has everything of the moon's youthful promise. The

---

1. *Atlas Coelestis seu Harmonia Macrocosmica* (The Celestial Atlas of Harmony) by Andreas Cellarius, Amsterdam, 1660.

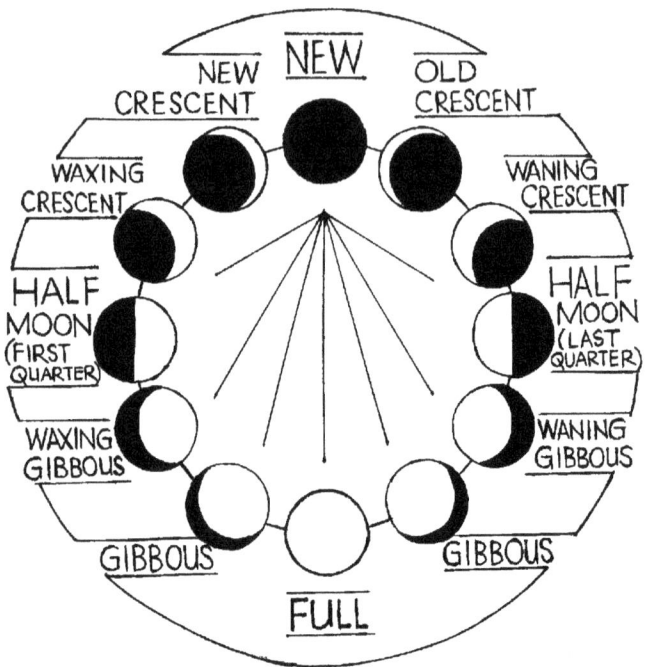

dark nights have gone. The first visible light in the bow of the crescent is symbolic of seeing our way ahead and productively focusing new plans into action.

At the other end of the lunar month the **Old Crescent** is the last of the twelve phases illustrating now the waning energy of the 12th harmonic, a semi-sextile or 30° in distance back to the New. From this moment the moon slips away into obscurity yet the aspect is positive, the realistic understanding is that all the work of the cycle has been done and one is now happily surrendering to the completion of its deep and vaster tides. Robert Blaschke[2] likened the number twelve and its Pisces-like signature to discipleship and the implication of giving oneself up to a greater ideal.

After the New Crescent comes the **Waxing Crescent**, fuller and brighter and now a whole 60° or sextile aspect from the New. The sextile is the aspect of opportunity and the growth is personally straightforward, harmonious and rapid. At the farther end of the month the **Waning Crescent** or the eleventh stage takes opportunities to reorient and merge rhythmically with the closing tide and divest the personal into the collective.

---

2. Robert P. Blaschke, *Astrology, A Language of Life. Volume II. Sabian Aspect Orbs*, Earthwalk, Oregon, 2000.

The **Half Moon** at both its **First** and **Last Quarters** is a phase found in the 8 and 12-fold lunar patterns and follows the general meaning of the square aspect at 90° to the New. This is a challenging phase when the forward thrust of the waxing moon hits the first real blocks in its path and must strive to overcome the difficulties encountered. At the opposite end the waning moon does not have the same energetic push but still meets the resistance of the 90° square aspect. Now it is required to slow up and let go and above all accept that the cycle has reached its harvest stage. Pushing ahead with something new will achieve little in this present cycle. In this last week of the lunar month the moon's light and energy will diminish fast.

The **Waxing Gibbous** stage is probably the most favoured of all the phases as this corresponds to a 120° trine aspect from the start of the cycle. The growth is now easy. After the uphill delays and re-routings at the preceding First Quarter stage the road is straight and carefree. We are heading for success, free-wheeling with the wind at our backs. The illumination of the Full Moon is in sight and we are hopeful that it will favour us. The **Waning Gibbous**, corresponding to the waning trine, allows that same optimistic spirit to flourish but now it will be after the illumination of Full Moon. It is a sigh of perceptive relief after accepting and reorienting to all that the Full Moon and the first half of the cycle has delivered.

The last waxing phase before the Full Moon proper is the **Gibbous** moon. This is 150° from the start; a quincunx aspect, an energy primarily of adjustment. It will flow rapidly into the moon at its totality, the clear illumination of the entire energy of this particular cycle. But it has one last chance to adjust to it. Will the Full Moon highlight success or shine like a searchlight on failure? The Gibbous moon makes you increasingly aware of what's coming, just as the Gibbous phase directly after the Full Moon makes you not only aware of what did happen but how you can adjust further to it, building on your success, or learning from your failure.

The two Gibbous phases 30° either side of the Full Moon create a sextile base to a Yod pattern that has its apex at the New. The degree opposite to the apex, sometimes called the release point of the Yod, would in this case be the moment of Full Moon itself, confirming that the entire meaning of any lunar cycle birthed at the New Moon is both realised and released at the Full Moon climax. This whole period of a little less than five days with the Gibbous phases encompassing the Full Moon at their core must therefore contain that fated quality of forces outside our control for which Yods are renowned.

The **Full Moon** is 180° from the New Moon with everything that the opposition aspect and full open light implies. The possibility of a full flowering, a complete knowing, the moon at the height of its powers, is underpinned with tension and the play of opposites. The Full Moon phase is naturally also found in the 8-fold division of the lunar cycle, as it must be in all divisions of the lunar cycle, as it is unquestionably the moon's most authoritative and forceful manifestation.

There are many other ways in which the moon's cycle can be split, including one with a circle of nearly 30 phases – a distinction for each day of the moon's monthly journey. With a separate name and meaning for each, these are the Mansions of the Moon, or the Nakshatras of Vedic astrology. In the early twentieth century W.B. Yeats' occult work *A Vision*,[3] gave his own version of 28 named phases and meanings:

> 'Twenty-and-eight the phases of the moon,
> The full and the moon's dark and all the crescents,
> Twenty-and-eight, and yet but six-and-twenty
> The cradles that a man must needs be rocked in;
> For there's no human life at the full or the dark…'
> [W.B. Yeats 1865-1939. From 'The Phases
> of the Moon' in *A Vision*.]

The Lunar Zodiac:
The Moon Mansions and the length of a lunar cycle

Nine times three equals 27, the usual number of moon mansions or Vedic nakshatras. Each lunar mansion measures 13°20' on the zodiac wheel, the number of degrees the moon travels in one day.

"But the moon doesn't always travel 13°20' degrees a day," I hear you cry, as you turn the pages of an ephemeris counting the moon's daily motion from one day to the next.

That's right, it doesn't. It varies from about 11°47' to 15°21'.[4]

13°20' is taken as an average when dividing 360° by 27. (13·33 recurring). And it doesn't take 27 days exactly to traverse the zodiac either. It's more like 27 and a third days.

---

3. W.B. Yeats, *A Vision*, Macmillan and Co Ltd, London, 1962. Originally printed privately in 1925.
4. And these are not hallowed figures. It would differ slightly each year and would differ according to the time of day you started counting.

This leads to another dilemma – Why are there 27 moon mansions in some systems and 28 in others, and 29 or 30 days if you count from one new moon to the next in an ephemeris? How many days *does* the moon take to travel its round?

Apart from the fact that some systems of moon mansions did not have equal divisions, the length of a lunar cycle in general is easy to get confused about. The moon takes 27·3 days approximately to return to its starting point in the zodiac. From 1° Aries to 1° Aries for example – a lunar return – takes just over 27 days. But it takes more like 29·5 days to travel from one new moon to the next new moon. (Because each new moon begins roughly 30 degrees on from the last one, and the sun needs to join in, and the sun is always moving forward too).

So a lunar return is *not* a lunar month.

The mansions of the moon are concerned with dividing the zodiac wheel into 27 or 28 separate mansions or houses. As mentioned above these are the divisions for each day (each twenty-four hours) of the moon's journey through the zodiac, and each mansion has a name and character of its own. For all intents and purposes it forms a lunar zodiac that exists separately yet still wedded to the solar one.

The mansions of the moon were once an important part of astrology almost universally, but their use died out in the west and remains neglected to the present day. In India however the astrology of lunar mansions, called the nakshatras, has continued in an unbroken tradition and is regarded as an essential component of Jyotish.

In Komilla Sutton's *The Essentials of Vedic Astrology*,[5] I had the privilege of illustrating the 27 nakshatras in a chapter describing their energies and I also produced a nakshatra wheel that combined the tropical and sidereal zodiacs, which I have reproduced again below. In Vedic astrology the first nakshatra, Ashwini, begins at zero degrees Aries of the sidereal zodiac, which is approximately 23 to 24 degrees different on the zodiac wheel to the western tropical zodiac. Those who use the tropical system of astrology, the normal western system, will find tropical zodiac measurements shown on the inner ring of the illustration. These can then be related directly to the Vedic nakshatras on the outer ring. The diagram can be used quite simply in

---

5. Komilla Sutton, *The Essentials of Vedic Astrology*, The Wessex Astrologer, Bournemouth, 1999.

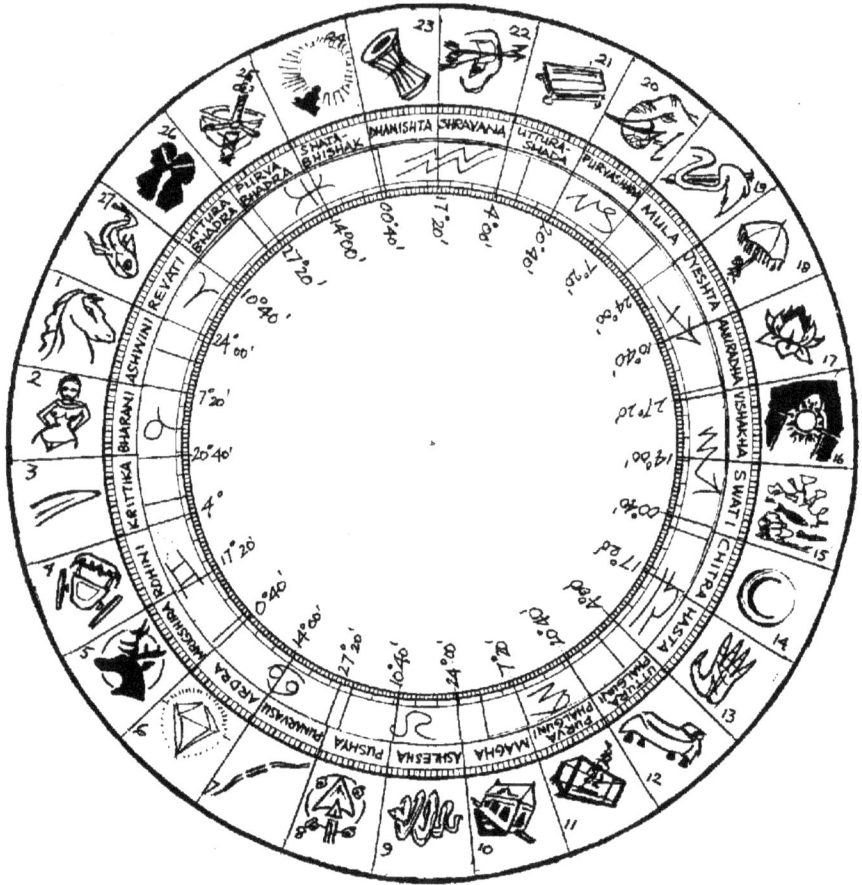

Vedic Nakshatras converted to tropical zodiac

both systems to find the nakshatra of your natal Moon (or any planet) and avoids the need to subtract or add degrees when converting the tropical or Vedic zodiacs into each other.

There are many available books of reference regarding the full meaning of the nakshatras and the activities favoured when the Moon is passing through each of them. I recommend a more detailed study than I can offer here if you are interested in this area. Below is a brief summary of the names, symbols and basic energies of each nakshatra synthesised and paraphrased from a variety of sources.[6]

---

6. My sources include: Komilla Sutton, *The Essentials of Vedic Astrology*, The Wessex Astrologer, Bournemouth, 1999.
Dennis M. Harness, *The Nakshatras*, Lotus, Wisconsin, 1999.

1. *Ashwini.* The Horse's Head. Good for new starts and beginnings. The light and dextrous path of a healer or shaman.

2. *Bharani.* The Yoni. (Female sex organ). Creation. Intensity and transformation. Understanding money and sexuality.

3. *Krittika.* A Razor or Knife. Ambitious, protecting, heroic and sharp. The warrior path.

4. *Rohini.* A Chariot. Beauty and passion. Good for weddings. Steady enjoyment of life.

5. *Mrigashira.* A Deer's Head. Gentleness in hunting and travelling. Romantic searching.

6. *Ardra.* A Diamond, Jewel or Tear Drop. Absorbing energy to become radiant. Disciplined physical activity.

7. *Punarvasu.* A Bow or Quiver of arrows. Abundance. Generosity. Travelling and returning with wisdom.

8. *Pushya.* A Flower, Arrow and Circle. (Also the Udder of a Cow). Nourishment. Lively ideas. Material and spiritual expansion.

9. *Ashlesha.* A Coiled Snake. Hypnotic and sexual. Uncovering secrets. Uncurling kundalini.

10. *Magha.* A Palanquin. (House on a pole). Majesty. Fame. Ancestral authority.

11. *Purva Phalguni.* A Bed and a Fire. Love and Desire. Relaxing enjoyment. Domesticity.

12. *Uttara Phalguni.* Four legs of a Bed. Love. Sexuality. Stability. Leadership.

13. *Hasta.* A Hand. Flexibility and skill. Craftsmanship. Dexterity. Serving and communicating.

Roeland de Loof and Martha Ijzerman, *The Hindu Lunar Zodiac. 27 Ways to Spiritual Growth*, (set of cards and explanatory book), Dirah Academie, Tilburg, The Netherlands, 1997.
Various articles and columns in *The Mountain Astrologer*, Cedar Ridge, California. Particularly Linda Johnsen's 'What's Your Vedic Sign?' June/July 1998, and the 4-part series 'Know Your Nakshatras', 2006-7. Also Tamiko Fischer's 'Nakshatras in Daily Astrology', April 2011.

14. *Chitra*. A Pearl. Charm. Artistry. Brightness and attractiveness. Opening the true self.

15. *Swati*. Coral. Restlessness. Independence. Adaptability. Material growth and gain.

16. *Vishakha*. The Archway. Heroic threshold to ambition and knowledge.

17. *Anuradha*. A Lotus. Instinctive. Blossoming in darkest corners. Revealing light.

18. *Jyeshta*. An Umbrella and an Ear-Ring. Seniority and status. Wealth and courage.

19. *Mula*. The Tail of a Lion or an Elephant Goad. Roots to power. Goading on the spiritual path. Overcoming fear.

20. *Purva Ashadha*. An Elephant's Tusk or a Fan. Popularity. Wisdom. Invincibility.

21. *Uttara Ashadha*. The Planks of a Bed. Wisdom through worldly happenings. Responsibility of karma.

22. *Shravana*. An Ear and an Arrow. Inner knowledge through listening and reflection. Speaking. Learning. Teaching.

23. *Dhanishta*. A Drum. Musical talent. Channelling other's rhythms. Prosperity.

24. *Shatabhishak*. A Circle. A Thousand-petalled Flower. Inner flowering. Cosmic unity. Unobtrusive, supportive energy.

25. *Purva Bhadrapadra*. A Sword. Warmth and courage. Fighting for causes.

26. *Uttara Bhadrapadra*. Two Heads. Cheerfulness. Humility and kindness. Partnership happiness.

27. *Revati*. The Fish. Releasing truths. Compassion. Fertility and procreation.

Incongruous Bookends:
The Magic Squares of the Moon and Saturn

The closest astrological planet to the earth is the Moon; the farthest (traditional) planet is Saturn, with the rest of the family strung out between. These two may seem incongruous bookends – Saturn is dry, the Moon is moist; Saturn is structure, the Moon is changeable; Saturn is a reaper, the Moon a sower; Saturn is a parent, the Moon a child; Saturn rules Capricorn, the Moon rules Cancer; Saturn has the shortest daily motion, the Moon has the maximum daily motion; Saturn's metal (lead) has the least conductivity, the Moon's metal (silver) has the most conductivity, and so on.

But they do have an astrological resonance in their cycles. 29 days is the Moon's cycle, 29 years is Saturn's, with the progressed Moon and transiting Saturn moving at about the same rate through the zodiac. As Saturn takes just slightly longer than the Moon to complete this cycle, the phenomenon is sometimes known as Saturn chasing the Moon. Saturn and the Moon also have some interesting tie-ups in their magic squares.

Magic Squares or Planetary Squares are the grid of numbers associated with each of the traditional seven planets. They survive today mainly as mathematical oddities for whichever way you add together the lines of numbers on the square (horizontally, vertically or diagonally) it always totals the same.

Originally these sets of numbers were regarded as extremely powerful, acting as a focus for each particular planetary energy and invoking the angel or guiding entity associated with it. They were (and are) the power grid of the planetary energy employed according to specific needs.

The simplest of the seven planetary squares is that of Saturn, used to attract patience and resilience, or to cool down overheated situations or reduce body temperatures and fevers:

$$\begin{array}{ccc} 4 & 9 & 2 \\ 3 & 5 & 7 \\ 8 & 1 & 6 \end{array}$$

There is something very solid and basic about the Saturn square which is composed of the first nine numbers and totals 15 whichever way you care to add it up; vertically, horizontally or diagonally.

A Chinese legend claims that this square was first discovered in the markings of a tortoise shell and formed the basis of the *I Ching*. The story is satisfying astrologically because slow-moving tortoises are ruled by Saturn, (and you could say by the Moon too, as they carry their home with them and are soft inside a hard shell – like a crab). In the Western world the employment of magic squares as talismans gradually died out with the rise of science but the particular arrangements of numbers I am using here are attributed to Agrippa, (1486-1535). These were later used by the Hermetic Order of the Golden Dawn in the late nineteenth/early twentieth century.

It has been speculated that the Seven Wonders of the World were representations of the seven planetary energies and may have incorporated magic squares in their groundplans – or deliberately combined a sequence of numbers in the way the structures were built. It is true that lists of the Seven Wonders of the World tend to change with fashion but according to Levi, the Saturn 'Wonder' was The Temple of Solomon on Mount Zion and the Moon 'Wonder' the Temple of Diana at Ephesus.

The masculine Solomon is the archetypal wise old man, a fitting Saturn figure. His great stone temple housed at its heart the 'Ark' of the Covenant – already we see a moon image incorporated. And its two pillars *Boaz* and *Jachin* we meet again in the lunar Tarot card of the High Priestess. [See further below].

The Temple of Artemis at Ephesus was a huge marble structure to the moon goddess, three times larger than the Parthenon and built according to legend on a site previously sacred to the Amazons. It is associated too with the 'Lady of Ephesus', that ancient many-breasted icon.

Back to the magic square of Saturn. Other than rotating it or mirroring it, there is only one way you can arrange the nine numbers in its square to produce the special effect and that is as shown. The numbers have to be in this relationship to each other. While the larger squares for the other planets have more options in this respect, Saturn is basic and unchanging. Saturn is therefore regarded to have four special numbers: 3, 9, 15, and 45.

3 is the number of columns in the square
9 is the number of cells or individual numbers in the square
15 is the sum of any line
45 is the sum of every number in the square added together

9 is therefore a prominent number (45 also adds up to 9) and this links the square of Saturn to the square of the Moon, which is nine times larger with 9 columns of 9 cells. The Moon square has 369 as the sum of any line and 3321 as the sum of all its numbers together. (All these numbers total 9).

Nine has always been a mysterious number mathematically. Multiplying any number by 9 gives a result that is immutably 9. (9 x 11 = 99. 9 + 9 = 18. 1 + 8 = 9). If you write down a telephone number, then write it again in reverse order, then subtract the smaller from the larger, the result will always total to 9. Nine is a self-perpetuating energy. Nine months or nine moons is the length of a pregnancy, the watery gestation prior to a birth. It is a number of ripening and completion. The ninth harmonic (the birth chart multiplied by nine) is the final goal we are reaching to manifest. The navamsha.

The Magic Square of the Moon
The Magic Square of the Moon is an odd-order square and the largest of the traditional seven planetary squares:

37	78	29	70	21	62	13	54	5
6	38	79	30	71	22	63	14	46
47	7	39	80	31	72	23	55	15
16	48	8	40	81	32	64	24	56
57	17	49	9	41	73	33	65	25
26	58	18	50	1	42	74	34	66
67	27	59	10	51	2	43	75	35
36	68	19	60	11	52	3	44	76
77	28	69	20	61	12	53	4	45

The basic numbers associated with this square are as follows:

9 – the number of columns in the square
81 – the number of cells (or individual numbers)
369 – the sum of any line
3321 – the sum of every number in the square added together.

The Square of the Moon is nine times larger than the Square of Saturn. There are alternative versions of the square of the Moon with different ways that the numbers can be arranged to produce the end result. Saturn's square being the simplest and most basic arrangement has no other way other than a mirror image to lay out its nine numbers but the Moon with its square of 81 individual numbers has other options.

If you join up the numbers in the Moon square; that is drawing a straight line with a ruler from 1 to 2 to 3 and so on to 81; the end result looks like this:

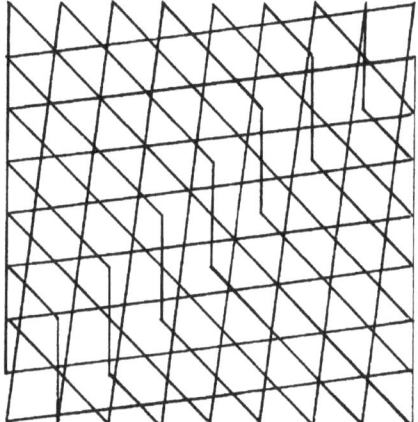

And when shaded looks almost like rippling water or moonlight through frosted glass:

The Square of the Moon can be used to enhance receptivity, memory, feeling; and remembering dreams. Or to regularise re-occurring cycles, menstrual or otherwise.

You can create your own personal seal on the Square of the Moon by first separately writing down your name (usually your full name at birth) and finding the numbers for each letter. The simplest method of doing this, without resorting to ancient Greek, Hebrew or other exotic alphabets, is to take the first letter of the alphabet (A) as 1, then B as 2, C as 3, continuing on to Z as 26. Underneath each letter of your name you assign a number to each. So JANE for example would be 10, 1, 14, 5.

You then connect these numbers together on the Moon's Square, beginning (in the example of JANE) at the number 10, then drawing a straight line from 10 to 1 on the Moon Square, then from 1 to 14, then 14 to 5. It is best to make these connections from and to the centres of the individual numbers on the Moon Square. With a full name this may run to a fair number of connections and eventually form a pattern or seal or sigil unique to the person involved. The final design, which is individual to you alone, can indicate a way of resonating, aligning and attuning personally to the Moon.

In India a 9x9 square is the basis for a traditional Hindu design (Vasta Purusha Mandala). In the western Tarot deck the Moon card is number 18.

The Tarot Card of the Moon

It's an old idea that the numbers involved in a conventional pack of 52 playing cards (not the Tarot deck) represent the solar year in miniature. There are 4 suits for the four seasons, 12 court cards for the twelve months, 52 cards for the number of weeks and, supposedly, 365 pips or spots on the face of the cards for the number of days in a year. (This last is highly imaginative but few bother to check it). There are some other lunar numbers not often mentioned like the total of non-court cards being 40 (the novile aspect; the forty weeks of pregnancy) plus those awkward rule-changers the Jokers representing the intercalary days or the bothersome calculation of joining the solar and the lunar year together.

The Tarot deck however is quite a different proposition.

A Tarot deck consists of 78 cards. It is comprised of 4 suits of 14 cards each – a total of 56 cards called the Minor Arcana, plus 22 special named cards called the Major Arcana. These numbers have little to do with the solar year – or the lunar, come to that. But there are cards labelled The Sun and The Moon amongst a host of other images. 'The Moon' is a major card, positioned eighteenth out of 22, or nine-elevenths of this major section. In a reading the cards are usually laid out in lines or squares or other mathematical arrangements previously decided. This clear-cut 'earthing' function is an important base for an intuitive reading of the meanings of the cards.

Modern Tarot cards are the property of everyone and come in an abundance of inventive designs, with Italian decks probably the most

imaginative of all. No longer are Tarot cards in the guarded silk-wrapped province of the overly mystical with their darkened rooms and borrowed rituals. Anyone can take their own approach and even create their own packs if they wish, yet certain conventions concerning card number 18, The Moon, are usually observed.

Over the past centuries a pictorial design has evolved that includes the following symbols:

The moon, commonly with a face, is at the top of the card shedding droplets of moisture to the earth below. Whether these moon-drips are tears, exemplifying the moon's connection with the emotions, or damp drops of dew, once believed to fall overnight from the moon, or the talismanic 'Tears of Isis' that flooded the Nile, or some kind of esoteric lunar rays, or even monthly menstrual blood; they fall between two towers or pillars and on to two wild dogs or wolves who are agitated and baying. From a pool at the foot of the picture a shelled water creature like a crayfish, lobster or crab, is rising into the moonlight. There is a winding road leading from this pool and between the two pillars into a barren undulating nightscape, reminiscent of lava plains or mounts of the moon on a palm.

The Moon is positioned numerically between the Tarot card of The Star, number 17, and the card of The Sun, number 19, both of which have bright images and fortunate meanings in a Tarot spread. But the Moon card is not so straightforward. Instead of a beautiful maiden as in The Star, or happy children as in The Sun, the Moon has rabid dogs and cold hard-shelled fish. There are usually no people in sight. In many Tarot decks the card of the Moon is unique among the twenty-two major trumps in having no human being depicted. It suggests that the moon's influence is instinctive and more on a par with the heightened senses of animals and nature. It rules a world that many of us humans feel uncomfortable to linger in, although we visit it nightly and it intrudes into a large part of our waking behaviour. The Moon card does not exhibit as dangerous an image as some of its near neighbours like The Tower of Destruction, but it is the fear of madness, of not being in total control of one's mind, the loss of ego, which presents the moon in this bleak uncanny landscape as disquieting.

However The Moon is not necessarily a negative card. It may point to the importance of dreams and intuition or of changing phases and uncertainty. It may refer to the timings of a lunar cycle.

Let us look at some of its symbols in turn. Firstly, what are the two pillars?

We also meet these two pillars, or at least a set of pillars, in Tarot card number 2, The High Priestess. Both The Moon and The High Priestess can be seen as feminine cards. The High Priestess, the 'Queen of the borrowed light' – as Waite calls her,[7] represents the more spiritual side of the earthy feminine displayed in card 3, The Empress.

---

7. A.E Waite, *The Pictorial Key to the Tarot*, Dover, New York, 2005. Originally published 1911.

To make the lunar connection clearer, The High Priestess often has a crescent moon at her feet or on her brow. In a Tarot reading she might depict a real person; whereas the Moon card, unlike in astrology, is not usually referring to a specific woman or mother.

The High Priestess holds secrets, and is ready to reveal them if approached correctly. The card depicts her on a throne facing us. We can see her eyes. She can be a purveyor of the meaning of the moon to us.

The two pillars of the High Priestess are usually clearer and more distinct than on the Moon card, where they are rough-hewn or featureless and always in the background.[8] The ornamented pillars of the High Priestess are coloured black and white. The pillar on the left is black, the pillar on the

---

8. In an early set of cards – the 'Cary' sheet c.1500 – they are believed to be lighthouses, marking a connection with the sea.

right is white. In the influential Rider-Waite Tarot pack[9] the black pillar has the letter 'B' while the white pillar the letter 'J', referring in this case to *Boaz* and *Jachin*, the two pillars of Solomon's temple and part of Masonic lodge design. But the dark and light pillars between which the Priestess sits can simply refer to the balance she holds between what is known and what is unknown, the conscious and the unconscious.

The High Priestess sits permanently between these solid sculpted poles of knowledge and wisdom. She can be a key to the moon.

If we think of the pillars as being the eyes of the Priestess looking out, then her left eye – her moon eye – is the white pillar, and her sun eye the black.

Secondly, what are the dogs?

It is believed that the two dogs, who may initially have been the hunting dogs of Diana, first appeared in the 'Marseilles' Tarot, a famous 17th century French set, and from the start they were distinguished by being painted in different colours. It is a feature not always observed in every Tarot pack that the two dog-like creatures howling at the moon should be differentiated either in colour or species. In the Rider-Waite deck one is a dog and one is a wolf, and they are certainly different colours. This is not a pair of identical Gemini twins and it seems important that the two animals must appear similar while different. In this respect they become perhaps like the High Priestess' pillars.

Wolves have a long association with the moon. They are known to howl at the full moon and folk legend has given them maternal instincts. There are stories of human infants reared by wolves in various cultures. Any amalgam of wolf and human has a lunar subtext; the werewolf who changes shape according to the moon's phases is a prime example.

Thirdly, what of the shellfish rising from the pool?

Whether it is a crayfish, lobster or crab this is one of the oldest symbols to appear on the Tarot card of the Moon. In some early sets the rising crab dominates the picture and it derives very simply from the Moon's rulership

---

9. Named after A.E Waite of the Golden Dawn who conceived it and William Rider who published it, this pack illustrated by Pamela Coleman Smith was first produced in London in 1909. It was special because all the 78 cards had individual pictures rather than just those of the 22 Major Arcana. The pack was republished widely in the English-speaking world when general interest in the Tarot blossomed from the 1960s onwards, and its designs still influence countless modern sets.

over the sign of Cancer the Crab. Astrology, which vastly predates the Tarot, would have influenced this obvious association of themes.

Should the Moon card depict a new, full or old moon?

Wolves bay when the moon is full and shellfish are said to move in accordance with it, but most packs, including Rider Waite, attempt to combine all the moon's phases within the sphere of the overhanging lunar globe. If there is a crescent here it should have its horns to the left, in the manner that we draw the Moon's glyph in astrology, as this is the growing waxing crescent of the new moon.

Similarly the High Priestess often has both a crescent moon at her feet and a full moon on her brow or a head-dress to convey that she encompasses all moon phases.

The Progressed Moon

Some see the effects of planetary transits as having a more outer effect while those of planetary progressions a more inner effect, though this doesn't always mean much as an outer event can affect you inwardly and vice versa. If both methods of astrological timing are to be employed perhaps we should favour progressions (secondary progressions) for the inner faster-moving planets Sun to Venus, which by transit are too fleeting to last long, and use transits for the outer slower-moving planets Jupiter to Pluto, which by secondary progression might hardly move in a lifetime. Mars can fall into both categories and Solar Arc directions are different again.

The fastest-moving planet of all is the Moon, hurtling around a zodiac chart at an average of just over one degree every two hours; moving an average of 13° in 24 hours, and through an entire sign in two and a bit days. It's never going to transit long enough over one zodiac degree to make more than a cursory flickering effect on a natal planet. Unless it's acting as a trigger for another longer slow-moving transit (by transiting the ascendant for instance while the slower-moving transit is labouring on), or it is joined to the Sun at a new or full moon, its passing mood on one zodiac degree is bound to be temporary and impermanent.

That's the Moon by transit. How different it is by progression.

Using year-for-a-day secondary progressions, the Moon will now hover over one zodiac degree for about a month. It will take over two years to traverse a zodiac sign entirely and about twenty-seven years to travel the whole zodiac. That's the straightforward progressed Moon. I also prefer to progress from lunation to lunation, 29 to 30 years by progression, using the method championed by Dane Rudhyar, elucidating the symbolic moon phase your life is passing through at any one time.

If on the night of your birth you had been able to observe the moon through your window, its phase in the sky (crescent, gibbous, full…) would be your lunar birth phase. But how many nights would you have to wait until the moon was new? An ephemeris answers the question so that if you were born on a waning moon you would only have a few days to wait. If you were born at full moon the wait would be about 14 days. And if you were born soon after a new moon you might have to wait as long as 28 or 29 days before the next new moon took place. Converting those counted days into years gives the symbolic measure of when certain new starts begin in your own life. If there was a new moon nine days after birth then at your age of nine an important new phase of life is set to begin. And much later at the age of approximately 39 years another fundamental new beginning would take place with the next progressed new moon. An average life span gives two or three progressed new moon beginnings each separated by 29-30 years dividing each life into vast periods of regeneration and growth. These commence at different ages for each person and depend on the lunar phase that the person came in with. As they begin at different degrees of the zodiac each time, unlike progressed moon returns, there is a different flavour or background to each successive cycle, determined by its proximity to a natal planet or fixed star or the pictorial degree symbol it falls upon. Should any of these progressed lunations be eclipses then the changes they denote are even stronger, as would progressed new moons falling close to zero degrees of Cardinal signs when the declinations would be closer to the extremities.

Tertiaries
Tertiary progression is another measure specifically devoted to the moon. Instead of relating each day in an ephemeris to one year of life, each day now becomes a lunar month. The Moon moves quicker by this method of progression and takes a little less than two solar months to traverse a zodiac sign (about 58/59 days). It will now hover over one zodiac degree for almost two days and take a little less than two years (approximately 1 year 11 months) to travel the whole zodiac.

Moon Hours

Traditionally each 24-hour day is divided into twelve planetary 'hours' of light and twelve planetary 'hours' of darkness. The daylight hours run from sunrise to sunset; the night-time hours run from sunset to the following dawn, so the length of an individual planetary hour differs according to the season of the year. In summer the twelve daylight hours are longer than the night-time hours with the reverse in winter. Reading the planetary hours from a horoscope is best achieved using the Placidus house system as each house will equal two planetary hours. This begins from the ascendant as dawn with the 12th house representing the first two hours of the day, the 11th house represents the following two hours and so on around to the seventh house and sunset. The night-time hours begin at the descendant with the sixth house representing the first two hours of the night, the fifth house the following two hours and so on. The house in which the Sun appears on a horoscope allows you to calculate roughly by eye the planetary hour governing the time of that particular chart.

As each planetary hour is ruled by one of the seven traditional planets in a continuing sequence that runs Sun-Venus-Mercury-Moon-Saturn-Jupiter-Mars, the hours that belong to the Moon will appear more on some days than others. A 24-hour day is understood to begin at sunrise, so that Monday for example does not begin at 00.00 hours on Monday morning and finish at 23.59 hours on Monday night; it begins at sunrise on Monday morning and finishes just before sunrise on Tuesday morning.

Each day of the week is ruled by the planetary hour that starts it, so that the first hour immediately after sunrise on Sunday is ruled by the Sun, the first hour after sunrise on Monday is ruled by the Moon and so on, with Tuesday ruled by Mars, Wednesday ruled by Mercury, Thursday by Jupiter, Friday by Venus and Saturday by Saturn.

Because the seven-planet sequence does not fit exactly into 24 hours (24 can not be divided equally by 7) some days – as already mentioned – will have more Moon hours than others. We can see from the table below that the days with the most Moon hours (four Moon hours in 24 hours) are Monday, Wednesday and Friday. The remaining days of the week, Sunday, Tuesday, Thursday and Saturday, have only three Moon hours.

Moon hours

Monday: 1st hour of the day, 8th hour of the day, 3rd hour of the night, 10th hour of the night.

Tuesday: 5th hour of the day, 12th hour of the day, 7th hour of the night.
Wednesday: 2nd hour of the day, 9th hour of the day, 4th hour of the night, 11th hour of the night.
Thursday: 6th hour of the day, 1st hour of the night, 8th hour of the night.
Friday: 3rd hour of the day, 10th hour of the day, 5th hour of the night, 12th hour of the night.
Saturday: 7th hour of the day, 2nd hour of the night, 9th hour of the night.
Sunday: 4th hour of the day, 11th hour of the day, 6th hour of the night.

The question is, as days of the week, are Monday, Wednesday and Friday more lunar? It is a fact that if you count through the entire planetary sequence you find that Monday, Wednesday and Friday have less Sun hours than Sunday, Tuesday and Thursday, so the answer must be Yes. (Saturn's day of Saturday has a deficiency of both Sun and Moon hours).

The Moon is very strong on Monday, the Moon's day, but also on Wednesday (Mercury's day) and on Friday (the day of Venus). It seems to confirm the yin or feminine energy of the Moon with a link to Venus and the hermaphroditic Mercury. Sun or yang energy is stronger on Sunday, the day of the Sun and on Tuesday (day of Mars) and Thursday (day of Jupiter). Of those days that have a deficiency of Moon hours (Sunday, Tuesday, Thursday, Saturday), Sunday and Tuesday may be the least lunar because Thursday and Saturday have at least two Moon hours in the night rather than the day. The Moon is therefore slightly more akin to Jupiter and Saturn than it is to the Sun and Mars.

What should you do in a Moon hour?

Read poems, write poems, watch a film, dream, meditate, stuff yourself with cake, have a hair cut ('*Best never be born than Sunday shorn*'), go sailing, stay at home, do some washing, be part of a crowd, wear silver, see your mother, draw up a family tree, be patriotic, have a bath, make a nest, be nostalgic, nurse an infant, swim, sit on a pumpkin, paint a picture, eat a cucumber, contact a female friend in the Netherlands, etc.

Things it's probably not worth doing in a moon hour:

Mend a watch, quantify the measure of parallax, sun bathe, demand attention, deck yourself in gold, audition for the lead part in *The Lion King*, bet on a certainty, make a laurel wreath, stand prominently in the centre of a traffic roundabout, itemise a catalogue, contact a male friend in France.

Some Zodiac Degrees with Moon Relevance
24 Aries: Gandanta Point

Gandanta is a Vedic term for the meeting of solar and lunar zodiacs. That is where the cusp of a moon mansion or nakshatra coincides with the cusp of a zodiac sign. To convert the tropical zodiac to sidereal we must subtract roughly 24 degrees so 24° Aries in the tropical zodiac becomes 0° Aries in the Vedic system and marks the beginning and end of the entire zodiac wheel (Pisces/Aries cusp). To be born with the Moon here is to experience a focus on its cycles of beginnings and endings and the depth and intensity of the soul's death and rebirth.

As the tropical zodiac moves further from the sidereal by approximately one degree every 70-odd years, the ayanamsha (the difference in degrees between the two zodiacs) gradually increases and currently stands at almost precisely 24°00'.[10] This means that if you are looking at charts with dates before the 21st century you should lessen this figure and a birth in 1940 for example would find a Moon at 23° Aries (tropical) more closely on this gandanta point.

24° Leo and 24° Sagittarius follow this same pattern and meaning as they are the only other meeting points between the Vedic solar and lunar zodiacs.

3 Taurus: The Moon's Exaltation

The Moon rules Cancer but there is no unanimous agreement as to which degree of Cancer is the most potent for the Moon's residence. The exaltations of the planets however do have specific zodiac degrees by tradition, and for the Moon this is the 3rd degree of Taurus (2°54' Taurus). The antiscion or parallel degree to this would be the 28th Leo (27°06' Leo). The common declination to both is 12°29' North.

The Moon as Final Dispositor

In whatever sign your Moon is placed, the planet that disposes of your Moon (the planet that rules the sign your Moon is in), tells you something else about your relationship with Luna. In fact it can tell you a great deal.

Any chart with the Moon in Cancer naturally displays a strong Moon because Cancer is ruled by the Moon, and if the Moon is the only planet on the chart to be in its own sign it may be the *final dispositor* of the chart and therefore even more significant.

---

10. As in the second decade of the 21st century using the Lahiri calculation (285AD as zero ayanamsha).

Using the dispositors of planets in signs rather than house cusps, and using modern planet rulerships, the method is to take each planet in turn and find the ruler of the sign in which it is placed. For example Saturn in Gemini would be 'disposed' by Mercury, the natural ruler of Gemini. Mercury would be the dispositor of Saturn. Moving then to Mercury on this same chart, its sign position would indicate the next dispositor. As Mercury is in Sagittarius (in the *Casablanca* example chart shown below) then Jupiter would be the dispositor of Mercury. And so on.

Sometimes, as in this particular case (the chart of *Casablanca*), everything will lead to a final dispositor that can go no further because that planet is in its own sign. Here it is the Moon in its own sign, suggesting that everything ends with the Moon; that the Moon has the final say; that this story will

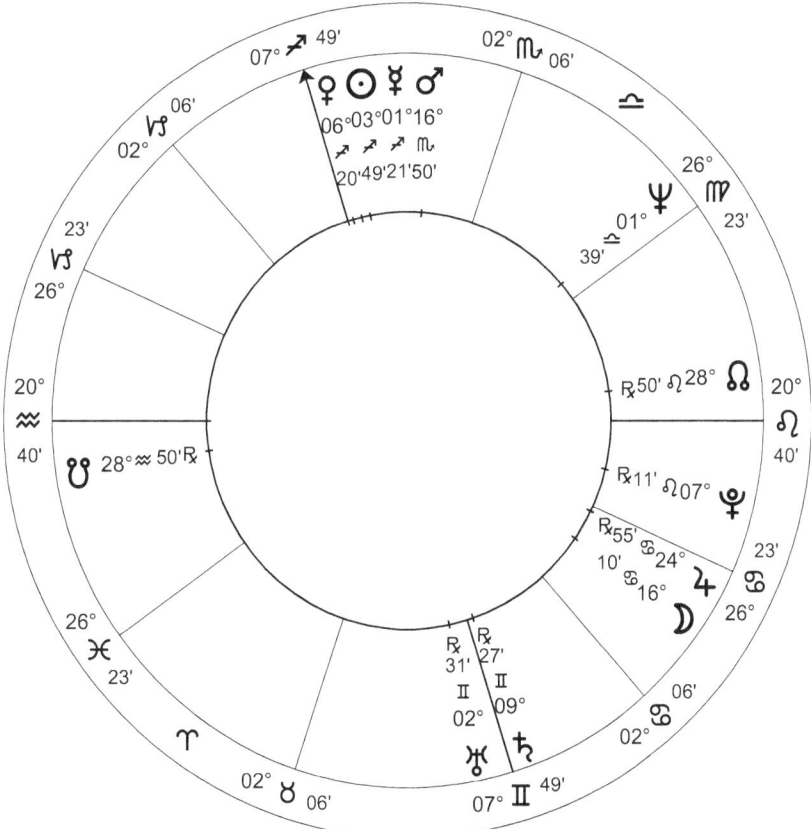

Casablanca World Premiere: Thursday November 26, 1942, at Hollywood Theater, New York City. Noon chart.

have an emotional pull on its audience and deal with issues of homeland and security.

Casablanca

Casablanca was a love story set in the cultural melting pot of Vichy-controlled French Morocco during the earlier part of World War II. The cinema film starred Humphrey Bogart and Ingrid Bergman and was successful in its time but no one could have foreseen the durability and classic status it would later achieve. It was probably the first Hollywood motion picture to gain a wide retrospective fan following – a wistful nostalgia for times gone by – and simply stated it is one of the most popular films ever made.

Venus conjunct the Sun in Sagittarius confirms that *Casablanca* is predominantly a love story in a faraway place. It revisits a romance between a Norwegian woman and an American man that takes place now in a French part of Africa in the hands of Germany. (And the Norwegian woman is married to a Czech). The climax takes place on a Moroccan airfield waiting for a plane to Portugal.

Venus and the Sun in Sagittarius, plus Mercury – the dialogue, are all disposed by Jupiter, the ruler of Sagittarius. Jupiter is situated in Cancer so is disposed by the Moon. The Moon is in Cancer. The trail ends here.

Mars, the movie's action, is in Scorpio. It is a case of do or die. Deals are underhand. Secrecy and intrigue are rife. This is Scorpionic black market country subject to darkly emotional factors. Mars is disposed by Pluto, Pluto in turn is disposed by the Sun, the Sun is disposed by Jupiter, Jupiter is disposed by the Moon. Everything comes back to the Moon.

The powerful Moon in its ruling position is not only conjunct the USA's 4th-of-July Sun but translates the title of the movie, *Casablanca*, into words associated with the Moon; literally 'white' 'house'. And more than all of this, the Cancer Moon describes the main environment of the story. When Humphrey Bogart surveys the local scene from the steps of *Rick's Cafe Americain*, we remember that the whole plot revolves around this place. It is both a casino (Sun in Sagittarius) and a Cancerian catering joint with a restaurant and bar, and it is the hero's moon-ruled home.

There is a special wistfulness to this Cancer Moon. It is the final dispositor of both Venus and Saturn, which are closely opposed on the chart. The haunting love-theme distils it, going by the title *As Time Goes By*. It is the story of an emotional rejection in the past that the main character has never forgotten; a love rekindled but now fraught with difficulties and responsibilities. The lovers are not morally free to love one other, and on a greater level it is a question of whether to put a personal relationship before the spreading global conflict. It is love versus honour (Venus opposite Saturn) on both counts drawn into a Moon in Cancer that is emotional and patriotic.

Casablanca ends with all the threads pulled into Luna, with 'not a dry eye in the house'. The Moon will have the last word, playing to the need of the emotionally-aroused audience. After the barely restrained passion of the lovers' parting scene, Bogart and the local police chief walk off together towards an unknown future. We understand they are going to join the Free French forces further down in Africa, but it was a message of reassurance to any in the free world that collaboration with the enemy would never be on the agenda of a true hero, or one who remained true to the best of their Moon.

Eclipses

Eclipses, both solar and lunar, have complex mathematical patterns. Each eclipse belongs to a lengthy series of over 70 or 80 eclipses called a Saros cycle, each eclipse moving on about ten zodiac degrees from the last. The eclipses in any one cycle are separated by a gap of 18-19 years so a complete Saros cycle can last well over a thousand years. There are dozens of different Saros cycles running all the time with their own starting dates and therefore their own characteristics and the 'total' eclipses in each will be towards the middle part of their cycle. Solar eclipses and lunar eclipses have separate cycles and it is well to remember that when a lunar eclipse follows on two weeks after a solar eclipse, apart from the fact that it is the full moon to a new moon in a monthly lunation cycle, this solar eclipse and lunar eclipse are from totally different Saros cycles and in that sense (and only in that sense) have no special relation to each other. A full moon is always born of a new moon but if they are both eclipses they come from different families. It adds weight to the thought that eclipses are rather individual new or full moons and don't behave as normal.

The Solar Eclipse: God is supplanted
A solar eclipse can only happen at a new moon and (usually) there are only a couple of solar eclipses out of the total of twelve to thirteen new moons in any year.

A solar eclipse is therefore an extra powerful new moon as the two great lights in the sky are aligned very precisely together, so closely that the disk

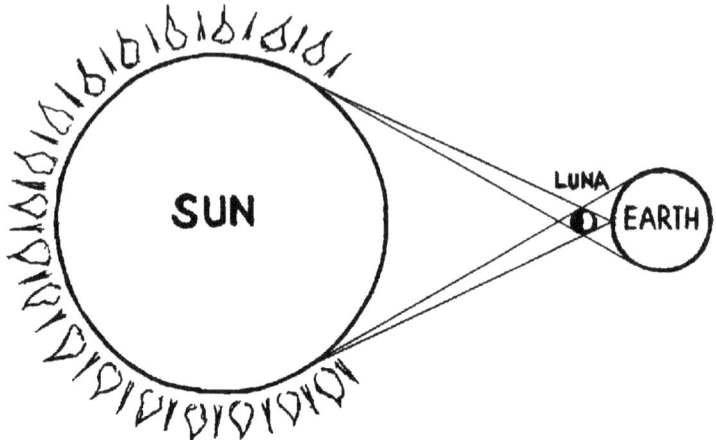

of the moon covers the disk of the sun. The moon has occulted the sun and completely hidden and blotted out its light and for a matter of minutes in one place the moon has total domination. The moon has come between us and the sun; the moon has temporarily divorced our relationship with the great life giver. The sun and moon may be at their closest union at a solar eclipse but it is the sun's light and power that has been abandoned and supplanted. It is the moon that has the upper hand and is extremely strong.

A Total Eclipse, as its name suggests, is one of the most awesome alignments in the heavens to affect the Earth. Not every eclipse is total. A partial eclipse does not completely block the sun and an annular eclipse, which we might call a Black Moon eclipse,[11] is similarly incomplete because the moon is at its farthest from the Earth. While there is likely to be only one total solar eclipse each year, if that, its exact location on the globe will differ. The eclipse is focused on a relatively small area about thirty miles in width,[12] which gradually moves across the earth in a band or corridor as the globe turns on its axis. As our world encompasses vast oceans many eclipses pass over the sea or can only be seen on remote islands or areas of land. Nevertheless astronomers and eclipse hunters regularly flock to these distant places to record and experience the total eclipse phenomenon. In today's computer age the path of an eclipse across the earth can be plotted

---

11. The annular eclipse can only happen when the moon is at its apogee, the Black Moon measurement. See more in the Black Moon section.
12. Stephen Hawking and Leonard Mlodinow, *The Grand Design*, Bantam, London, 2010.

instantly and accurately giving clues to an astrologer of the countries and land areas that might be most affected by the meaning of that particular eclipse. The pre-natal eclipse, either solar or lunar, (the last eclipse to take place immediately prior to one's birth), is also regarded as a clue to evolutionary life lessons, with the zodiac degree of this eclipse remaining potent throughout life. Geographically the pre-natal eclipse path may highlight particular land areas that may be relevant during a person's lifetime.

Eclipses affect the earth's magnetic current and on earth under the path of an eclipse all nature feels that the sun has abandoned it. Birds stop singing, animals settle down as if for the night and nocturnal animals stir themselves to life. With darkness overcoming light it is easy to see why doom-laden predictions could be hung on to all solar eclipses and why they were feared historically. If the sun disappears from the sky it is catastrophic for life on earth; the physical world is out of control.

By astrological tradition not every eclipse was necessarily regarded as bad. It portended a change if affecting an important part of a horoscope, as a powerful new moon would, but its nature depended on the planetary ruler of the sign that held it and fixed stars and planets that might conjunct it, and various other rules and maxims as likely to be positive as negative. Its effect on a chart was often believed delayed, sometimes prefiguring a crisis by many months, and while different time formulas for this have been set down it has to be said that none can be taken as 100 per cent accurate. The delay effect however has relevance when we consider the symbolism of the moon blocking the sun as the past blocking the present.

Above all at the solar eclipse the moon is in command, the Goddess is stronger than the God. The moon is about to begin its monthly cycle with added power attached – perhaps powerful memory attached. The past invades the present.

Eclipses have an association with the moon's nodes, the dragon's head and tail that swallow the sun or moon or the two wolves in Norse mythology who chase and catch the Lights. The sun and moon will be close to one of the nodes in longitude at an eclipse though not necessarily exact, but the moon's *latitude* must also be small for an eclipse to occur, in other words the moon must be crossing the ecliptic plane at the time of the lunation. The moon's nodes move backward though the zodiac taking about a year and seven months in each sign (18-19 years for the complete nodal cycle) but each year there are two eclipse seasons when the sun is roughly conjunct the north or the south node and in which new and full moons can become solar and lunar eclipses.

Pink Floyd and the 1999 Solar Eclipse
Musically speaking not many hit tunes have had planetary titles, although of all the heavenly objects the moon is easily the songwriter's favourite. This long predates rock and roll and is more the province of megaphone-wielding crooners in stripy blazers before and between the two world wars. To croon or spoon beneath the moon in June was about the size of it then.

The influence of the moon, according to the songs of a hundred years ago, was primarily for lovers only; its irrationality no more than a spell of love for those who sat beneath it. By this reasoning if you were not a young thing and not in love then the moon had little effect. The rock and roll generation also made anthems of being a teenager in love, yet that out-dated moon-in-June stuff was not their style. It was more the moon's emotionally passionate and dark side that instinctively hooked the rock musicians, including one of the best-selling albums of the twentieth century: Pink Floyd's *The Dark Side of the Moon*.

Luna can be a dangerous muse and those teenage dirge-hymns about motorbikers roaring off into the night and into their deaths after hot-blooded misunderstandings with their girlfriends, illustrated this leaning towards the dark Seductress of the Night. It's a safe bet that tangles of emotional irrationality, domestic upset and road accident, could be blamed on the moon.

Back in the 1950s an event that was coincident with the birth of rock music was the Space Race. In 1957 the Soviet Union shot up its *Sputnik* and suddenly Luna was not the only satellite in orbit. Man-made moons proliferated as fast as the growing music industry. Perhaps the greatest psychological step in lunar terms was a decade later when the far side of the moon itself was seen. As rock music plunged into a heavy, progressive era, the dark side of the moon was exposed to all, no longer a covert mystery.

In astrological/psychological terms this was as huge a step as the soon to follow small-step/giant-leap lunar footprint. By viewing its long-hidden dark side, the human race had made conscious the moon's secrets. Many things once taken as moon-ruled now shifted to Pluto's domain and much of Hecate's business went down to Erishkigal in the Underworld. It could be argued that rock and roll itself had a strong Pluto-Scorpionic flavour from its folk etymology. The term rock 'n' roll originally being a euphemism for sex.

Pink Floyd were one of the bands who came to prominence when the moon showed its hooded face in the late Sixties, and in an odd way they would always be associated with it. Unusually for a pop group at that time, they were commissioned by the BBC to compose music for the actual 1969 Moon Landing. Then, just a few years later in 1973, they released their most famous album picturing a prism of light on total black entitled *The Dark Side of the Moon*. Some critics saw it as an album about depression but I'm not sure that is how most people took it. *The Dark Side of the Moon* was a sensory audio experience in the days before video, when you closed your eyes, sank your head between two speakers and submerged yourself into an exciting and limitless void.

'Soul' and 'Blues' are both examples of moon mood music and Pink Floyd named themselves after two blues artists: Pink Anderson and Floyd Council. The main creative force in Pink Floyd, after the departure of Syd Barrett in 1970, was Roger Waters who had much to do with the Dark Moon album. The aforementioned Syd Barrett, who was born three days after a solar eclipse in Capricorn,[13] was a Sixties' casualty who met the Moon's dark face early and never fully returned to tell the tale. His experimentation with mind-expanding drugs may have helped produce one of Britain's first psychedelic LPs, *The Piper at the Gates of Dawn* in 1967, but the price was a loss of his grip on reality. One might say a loss of his Capricorn Sun. As if to confirm the lost Sun symbolism, Barrett's own father had died while he (Barrett) was still a teenager. Oddly too the band's second album in 1968 featured a track titled *Set the Controls for the Heart of the Sun*, as if they were still trying to find it. Roger Waters, though not born on an eclipse,[14] was similarly fatherless. His father had been killed in World War 2.

In August 1999 the end-of-millennium total solar eclipse was also tied to Pink Floyd and their music. A btnet website sought permission to play a track from *The Dark Side of the Moon* called *Eclipse* while the skies darkened and the actual eclipse took place. It was claimed at the time that both the track titled *Eclipse* and the cosmic eclipse itself were to last exactly two minutes and six seconds. In retrospect that may have been fanciful but the fact remains that Pink Floyd have always been connected to the moon.

Leo 18 was the degree of the great 1999 Solar Eclipse with its fixed grand cross and heavy millennium and Nostradamian overtones. It is tempting to see this Great Eclipse, part of the Saros cycle '1 North' or '145', as having

---

13. Syd Barrett born 6 January 1946.
14. Roger Waters born 6 September 1943.

a bearing on that world-changing event just two years later, the September 11th Twin Towers attack in New York City, (the 'terror from the skies' that Nostradamus had dated as 1999). Placing the charts of these two happenings side by side, only one degree is common to both: Leo 18. In 1999 this was the degree of both Sun and Moon at eclipse (18° 21' Leo), in 2001 it was the degree of the chart-ruler of that calamitous event, Venus (18° 24' Leo).

Although the 9/11 disaster had global consequences it was essentially traumatic for the people of the United States and the 1999 eclipse was not visible from the United States. The path of the 1999 eclipse moved across Europe touching England, France, Belgium, Luxembourg, Germany, Austria, Slovenia, Croatia, Hungary, Yugoslavia, Romania, Bulgaria and the Black Sea; then continued on through Turkey, Iran, Pakistan and India. This, plus the fact that the Sabian symbol for the 19th degree of Leo describes a party on a houseboat and is by no stretch of the imagination doom-laden, suggests a more visual connection with Europe and its immediate future. The houseboat party has a theme of different people gathering together convivially, and whether or not you agree with the word 'convivially', I think this is more concerned with the expansion of the European Union in the first years of the new century with many Eastern European nations joining the original fifteen members, plus the introduction of the common currency the Euro. (Especially as the ascendant of the European Union, formed 1st November 1993, was 17+° Leo – only one degree away).

But the eclipse had travelled right across the Atlantic, commencing its path in the ocean just south of Nova Scotia, which on the scale of a world map is not far out from New York, (it appears to point to New York), and other sets of degree symbols for 18+° Leo give images more akin to the New York and Washington 9/11 attack. Sepharial's translation of La Volesfera describes: '*A man running in the face of a strong wind, but making little headway. His garments fly in tatters behind him*'.[15] The Victorian Charubel sees '*A star surrounded by many rings*',[16] which admittedly could just be a vague reference to any eclipse. But the 21st century Fairchild symbols describe '*An alien king with an oval shaped head reading from a scroll to a young man who soaks up his every word*',[17] which sounds like a religious indoctrination from an 'alien' mastermind.

---

15. Collected in Alan Leo's *The Degree of the Zodiac Symbolised*, Fowler, London, 1898.
16. Also in the above-named collection.
17. Alice Kashuba, *The Fairchild Symbols*, Astroviews, Florida, 2003. (Images received in the year 2000.)

Because its path of totality moved across so many highly populated countries the 1999 solar eclipse is claimed to be the highest viewed eclipse in world history.

The Lunar Eclipse
A goddess in purdah

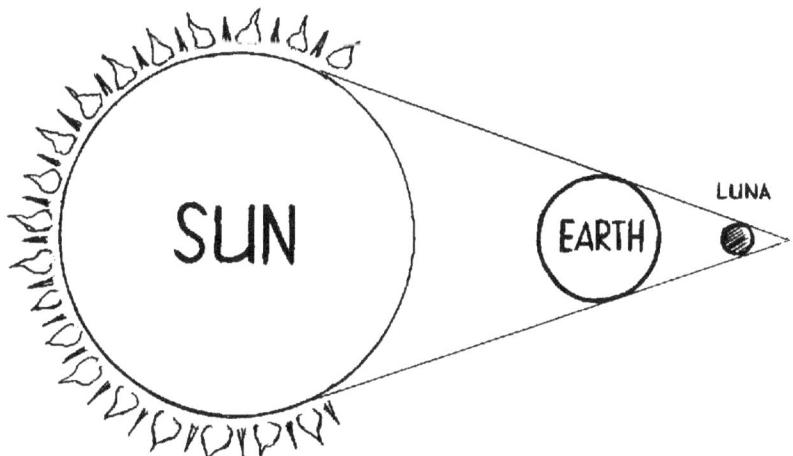

A lunar eclipse can only happen at a full moon and (usually) there are only a couple of lunar eclipses out of the total of twelve to thirteen full moons in any year.

A lunar eclipse is therefore an extra powerful full moon as it is aligned very precisely opposite the Sun. But at the same time an eclipse – to the inhabitants of Earth – means what it says; something has been eclipsed or shaded out, and in this case it is the moon.

The whole thing is a very different proposition to the *solar* eclipse when the sun disappears, because in that case the moon, being in the same degree as the sun, has the upper hand. For one long and dramatic moment it occults the sun and makes night-time of daytime. It may be a kind of marriage but the moon is much stronger than the sun at a solar eclipse.

But in a *lunar* eclipse the sun is not in the same degree as the moon; it is opposite to it right over on the other side of the world. It is the earth's shadow now falling on the moon that dims the lunar surface, darkening the 'pale fire she snatches from the sun'. So while the sun is even less powerful at a lunar eclipse than it is at any full moon, the moon itself is in a shadow state.

The essential difference between a regular full moon and a full moon eclipse is that one is shining and the other isn't. An ordinary full moon is a time of lunar illumination, with everything that that implies to us for good or ill. A lunar eclipse, still a full moon in essence, has drawn its curtains on us, making it much harder to know. We understand it is a full moon but it doesn't want to be seen as one. By showing us no whisper of shining silver light it is dully aping the appearance of an exact new moon in the sky. It is a full moon dressed as a new moon. A powerful mother hiding her experience of womanhood in the guise of a maid. A radiant goddess in purdah. Maybe even mutton dressed as lamb.

All of which could imply that the Moon is uncomfortable at a lunar eclipse because she is either deliberately hiding her true self or she is not totally sure who she is meant to be, and we (the earth) may be held to blame for it. One meaning often attached to a lunar eclipse is that the emotions are out of control. Certainly it leaves us all a little confused including the properties of the earth itself – eclipses are not considered favourable times for planting or sowing according to gardening manuals. It is probably best that we don't push the Queen of the Night for answers or look at her too closely; she may be emotionally touchy at this time, unable to contact her roots in the past while she is trying to find a face for the present. A problem might arise if the degree of the lunar eclipse is on an important part of our chart, like the Sun, Moon, Ascendant, Chart-ruler etc, which means that she's right in our line of sight and we can't avoid looking at her. We might expect she will turn away abruptly and leave us feeling empty, but we cannot be absolutely sure what she will do. She is not sure which phase she's in.

To be born on a lunar eclipse suggests that the Moon is still stronger than the Sun, and as the lights are in opposite signs the Moon's sign is more dominant than the Sun's. Thus it may take longer for that person to truly find themselves as the qualities of their Sun, the heart of the chart, has been experienced in a secondary position to the Moon. Much of who they feel they are may be the opposite of their Sun sign, (much of who they are *is* the opposite of their Sun sign), but as in any opposition between two planets one is favoured or owned more naturally than the other. An opposition always strives to be a conjunction, to unite the strength of its two energies and become them both, but in the process it can swing between the diametrical extremes, easily projecting outwardly the non-favoured planet on to other people and situations. When the 'favoured' planet is the Moon in eclipse, the search for 'who am I?' is intensified.

Any opposition between planets has the creative dynamic associated with the Second Harmonic. The continual facing off of two energies eventually produces a will to incorporate parts of both in a complementary fashion and in the process strengthens both rather than weakening one at the expense of the other. Thus the Moon in eclipse paradoxically finds itself or understands better who it is by incorporating some of the qualities of the Sun that it is hidden from.

In the sky the lunar eclipse has quite a lengthy duration, several hours compared to the few minutes of the more impatient solar eclipse, and it has no specific path across the Earth, it can be seen from anywhere. So if the moon wishes to hide her most open and brightest face it is not going to be easy to identify her meaning. Traditionally a lunar eclipse was not always an outright evil omen, rather a reduction in strength or a disappointment in outcome that might last a lunar month for each hour of the eclipse, or that may be activated later when another planet transited the zodiacal degree of the eclipse.

But it does bring an opportunity to escape from the past.

Bad Moon Rising:
The lunar eclipse that marked the end of the Swinging Sixties
A Bad Moon Rising with turmoil and disaster in its wake were amongst the warning lines of a hit record in the late summer of 1969 (*Bad Moon Rising* by Creedence Clearwater Revival) that coincided with a suitably powerful lunar eclipse on September 25th at 2°Aries 35'.[18]

Only weeks after the famous Woodstock music festival and in a year when tuneful wisps of the dawning of the Age of Aquarius still hung in the air like magic smoke, *Bad Moon Rising* foresaw an altogether bleaker future. Just like the doomsday language of medieval astrologers forecasting a lunar eclipse of plague and pestilence, the hit record spoke of hurricanes, floods and earthquakes. While many rock songs have a hard uncompromising edge, *Bad Moon Rising* with its simple foot-tapping spirit-lifting sound betrayed an uncharacteristic prophecy of ecological doom.

Viewed in retrospect this eclipse may have marked the finale of that decade-long party called the Sixties that you only remember if you weren't really there, as they used to say. But the Charles Manson murders in

---

18. In the United Kingdom *Bad Moon Rising* first reached Number One on the best-selling records' chart during the week ending 20 September 1969. Source: *500 Number One Hits*, Guinness Superlatives, Enfield, 1982.

California and the fatal knifing of an audience member at a Rolling Stones concert finally put the lid on the Sixties for many people. Sir Kenneth Clark's immensely popular TV series *Civilisation* ended its view of history by coming up to date in 1969 with its genial host somewhat perplexed.

"I am completely baffled by what is taking place today", he admitted with endearing honesty, and brought in a quote from Yeats about things falling apart and the centre not being able to hold[19] to describe his loss of confidence. It sounded eerily like a more 'civilised' version of Bad Moon Rising.

The lunar eclipse in September 1969 was noticeably potent because both the Libran Sun and the Aries Moon exactly squared Mars in Capricorn producing a forceful Cardinal T-square. It was part of a relatively young Saros series of eclipses that began back in 1843[20] with a parent lunar eclipse on 11 July 1843 at 18°41' Capricorn.[21] This Capricorn/Saturnian flavour joined with the Bad-Moon-in-Aries-square-Mars-in-Capricorn energy to give out a more aggressive down-to-earth realism set to jar the swinging decade. There is no doubt that this eclipse was powerful for the full moon was spread across the celestial equator, while Mars was placed at a declination of 26°06' south of the equator, beyond the ecliptic and more extreme than usual.

In the music world this eclipse (2°35 Aries/Libra) fell on or close to the natal Neptune of a wide spectrum of contemporary singers and musicians born between 1942 and 1945, and millions of young people who might have experienced the eclipse of a collective dream (Neptune) in 1969.

In one sense the lyrics of *Bad Moon Rising* were at least an eclipse cycle (nineteen years) ahead of their time. The following decade, the 1970s, sometimes called the 'cynical' or 'sobering' Seventies after the highs of the preceding years, were not so much full of Biblical rage and ruin as were the 1980s and early 90s when Pluto rampaged through Scorpio. An eye taken for an eye – the gist of the last line of the last verse – is vengefully Scorpionic and not much reminiscent of the Sixties' style of revolution nor the Seventies' following mood. If the song itself was prophetic, foreseeing the extreme weather patterns that would begin to make headlines around

---

19. From *The Second Coming* by William Butler Yeats. First published 1919. Also quoted in the book *Civilisation* by Kenneth Clark, BBC, London, 1969.
20. Young because this series does not peak until 2492 and will not end until well beyond the year 3000.
21. Information kindly supplied by Sylvia Jean Smith of Ottawa, Canada.

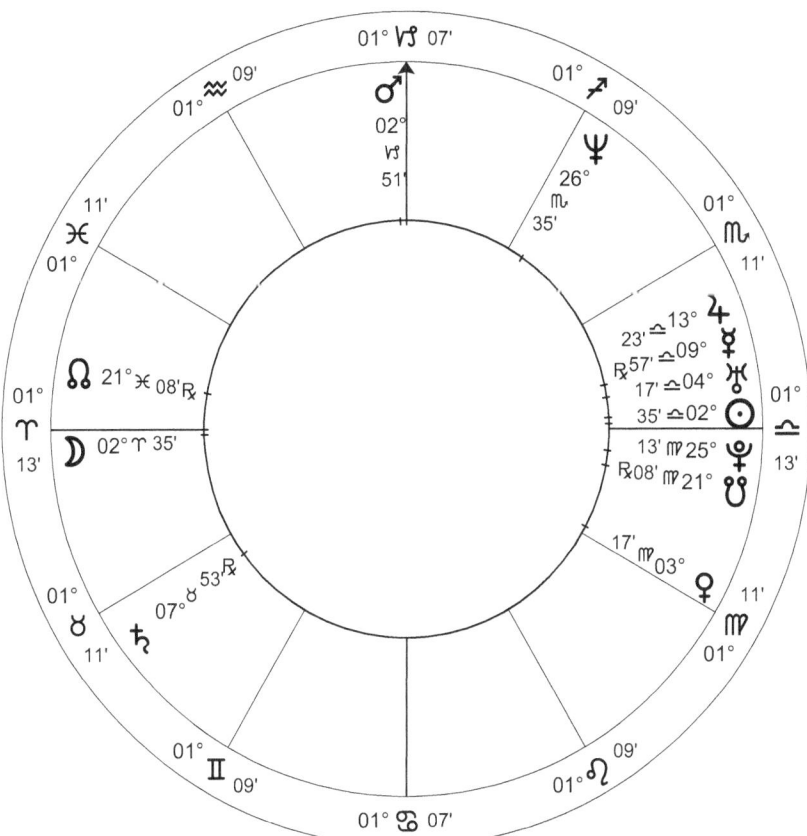

Bad Moon Rising. 25 September 1969. 20.21 GMT. Salvador, Brazil. [The chart is set symbolically for one of the places on earth that the Moon would have been rising at the moment of eclipse. Salvador in Bahia is the old colonial capital of Brazil, a country that holds the Amazon rain forests seen as vital to the ecological balance of our planet.]

the world in the 1980s and 90s with the discovery of the Ozone hole and Global Warming, perhaps the eclipse also was.

But at the time of the eclipse in 1969 it was the war in Vietnam that was the burning political issue. The degree of the eclipse (2° 35' Aries) was conjunct the natal Moon of North and South Vietnam (2° 43' Aries)[22] and

---

22. Nicholas Campion *The Book of World Horoscopes*, Aquarian, Wellingborough, 1988. The two separate states of North and South Vietnam were created by The Geneva Accord, signed at 3.45am July 21, 1954.

coincided with a change of leadership in North Vietnam after the death of Ho Chi Minh almost to the day. After reluctantly waging a war that was out of place in an anti-war decade, the first young American troops began to pull out of Vietnam in the summer of 1969, although the bombing by air was escalated. Three weeks after the eclipse the Vietnam Moratorium was held in the USA with millions of people protesting against the continuing involvement. It was the biggest anti-war demonstration ever held in the States. Astrocartography shows that, at the moment of the eclipse, Neptune was culminating at Washington, while it was anti-culminating at Hanoi, both places being exactly half a world away. The eclipse was dissolving their ties. And while the confusion of an indecisive war in South East Asia was being brought to a head, the Sixties themselves were soon to go out with a bang. Bouncing happily from the walls of a thousand discotheques across the planet in the dying summer days of 1969 were merry doom-laden words reminding us all to be prepared to die.

Many pop songs of the late 1960s were poetic and visionary, giving glimpses of a new age. Yet alongside the musical tapestries flowing from the hands of some progressive bands, Creedence Clearwater Revival were a down-to-earth, singles-market rock band whose message in this instance was so different and in some ways correct. And who knows, the fact that there wasn't a global disaster of famine, plague and everything else around the eclipse of September 1969 may be due to the international broadcasting of *Bad Moon Rising* acting as the benign lightning rod.

THE POETIC MOON

'What is there in thee, Moon!
That thou shouldst move my heart so potently?'
[John Keats 1795-1821.
From *Endymion*]

Moonscapes

The Moon in Art

As Professor Paul Murdin of the Institute of Astronomy in Cambridge pointed out in a lecture on the nineteenth century landscape painter Samuel Palmer (who was fond of depicting the moon in his pictures), although the moon has no atmosphere in a scientific sense, in art its very purpose is to generate atmosphere.[1] Of all the planets and heavenly objects that wheel above us, the moon is the artist's and poet's favourite because it can encompass every changing mood.

Take Max Beerbohm's tongue-in-cheek tale of the Oxford femme fatale *Zulieka Dobson* (first published in 1911). Into an all-male scene of young gentleman scholars in boaters and gowns comes Zuleika Dobson – a creature to die for. This doe-eyed little conjuror with 'no waist to speak of' can only love those who don't love her; which more or less excludes everyone. But my point in mentioning this book is that in one flowery passage Beerbohm compares the moon to a 'gardenia in the night's buttonhole', then interrupts himself and wonders why writers are always likening the moon to something else entirely, usually that bears no resemblance to it in any way. So he starts his paragraph again: 'The moon, looking like nothing whatsoever but herself…' He is parodying all kinds of Edwardian conventions in his writing but the point we can take here is that the moon is the most obliging symbol of comparison not only to different moods but to different objects. No other planet or star, including the sun, can be poetically accepted as so many different things.

As mentioned in the opening paragraph the visionary painter Samuel Palmer, born 27 January 1805 in London (time not known), actually painted the moon in many of his pastoral landscapes. To see the moon in the sky was unusual for paintings of rural scenes and it remains a defining feature of Palmer's work. Samuel Palmer was a child prodigy, a sensitive soul who first exhibited in the National Gallery at the age of fourteen. His paintings from his early twenties are now recognised as masterpieces, modern, mystical and uplifting, although they were not particularly popular at the time and it was his more mundane work that pleased the public.

---

1. Paul Murdin, 'Representing the Moon', The Inspiration of Astronomical Phenomena, Magdalene College, Oxford. 3-9 August 2003. Collected in *Culture and Cosmos*, Vol. 8, Nos 1 and 2, Bristol 2005.

Palmer joined a small band of artists called 'The Ancients' (1825-35), who had a similar poetic outlook to William Blake, who also met with them regularly. Others in the group included Edward Calvert, John Linnell and the artist/astrologer John Varley. The Ancients were predecessors in spirit to the artistic brotherhood of the Pre-Raphaelites in the latter half of the nineteenth century.

Where was Palmer's own natal Moon? It was either late Sagittarius or early Capricorn, depending on the time of day he was born, but what marks it out as unusual was its beyond the ecliptic, out-of-bounds status, deep in southern declination. In fact every planet on Palmer's chart was in south declination apart from Mars. But the Moon is the only out-of-bounder, the most intense, the most acutely sensitive, the most intuitive, visionary and psychic.[2]

The Moon as Muse

Everyone has a muse and in the case of gifted portrait painters we can sometimes see what this frustrating/amazing heavenly/devilish being looks like. The Pre-Raphaelite artist Dante Gabriel Rossetti (1828-1882) is a good example; he obsessively painted the same 'blessed damozel' all his life, though the models or sitters on whom this divine female was projected sometimes differed. Suffering for one's art is a pretentious phrase but the artist's muse is not always renown for being a reliable or comfortable creature. She's as inconstant as the moon. The writer Norman Mailer referred to his as 'the bitch'.[3]

Rossetti was born in the moon's darkest phase with Moon in the 12th house, exalted and opposite Jupiter.[4] The seven stars of the Pleiades were rising. In his poem *The Blessed Damozel* he describes the female of his inspiration thus:

> '…Her eyes were deeper than the depth
> Of waters stilled at even;
> She had three lilies in her hand,
> And the stars in her hair were seven.'

---

2. For more on the Moon out of bounds and declination generally see *Declination in Astrology: The Steps of the Sun* by Paul F. Newman, The Wessex Astrologer, Bournemouth, 2006.
3. *The Spooky Art*, Norman Mailer, Little Brown, London, 2003.
4. Dante Gabriel Rossetti: 12 May 1828, 04.30am, London, England. Source: Fowler's Compendium and AA data bank.

Norman Mailer was born on the day of a full moon.[5] In his 1973 book on Marilyn Monroe, *Marilyn. A Biography*, he invented the term 'factoids' and suggested that sometimes anecdotes and false media 'truths' give a better picture of the feel of an actor's life and personality than straight hard facts. It's a good example of that special kind of illumination and understanding that the full moon brings.

Another painter whose chart clearly speaks his muse is the surrealist Salvador Dali. 'Surrealism' as a movement attempted to take inspiration directly from dreams and the unconscious, creating in its art bizarre landscapes of the mind that needed no conscious explanation. It's pretty clear that we are in the province of the moon here.

Salvador Dali's natal Moon was at 2° Aries, conjunct the midheaven, square Neptune, conjunct the South Lunar Node. This is a moon that in one sense has only just entered the great panorama of life, as it is two degrees from the zodiac's beginning and 28 *minutes* from crossing the ingress equator (declination 00°28' north). It's all young and new as if woken from sleep into the morning light and still carrying the dreamworld with it. It is the ascendant ruler and it is angular, on the midheaven, forcing its moods and vitality directly into the public eye – a deliberately 'loony' public image. The Moon is in Aries, angular and chart-ruler – it's an open moon in every way and it completely dominates Dali's chart.

Yet it is also a moon that is rooted in older memories. It is an old moon by phase, and subject to the past energy of the South Lunar Node.

Angular planets often manifest as 'real', especially on ascendant or midheaven, and Dali did find this moon in a real woman. His beloved Gala (born Elena Dmitrievna Diakonova, 7 September 1894, Kazan, Russia) dominated his life. She was his world, his everything. He identified with her totally, even signing some of his paintings in her name. They first met in 1929; he was 25 she 34 and the wife of another artist. They soon became lovers and were inseparable, marrying five years later. He recognised Gala as the unknown woman he had always painted, usually from the back, and he continued to paint her (from front and back) for the rest of his life. In his 1952 painting *Galatea of the Spheres* Gala's face is an explosion of many globes. It almost appears that she is comprised of multiple moons. She may partly have been a mother figure to him, being ten years older and helping to manage his business affairs (Moon on midheaven), although by all accounts

---

5. Norman Mailer: 31 January 1923, 09.05am, Long Branch, New Jersey, USA. Source: Lois Rodden.

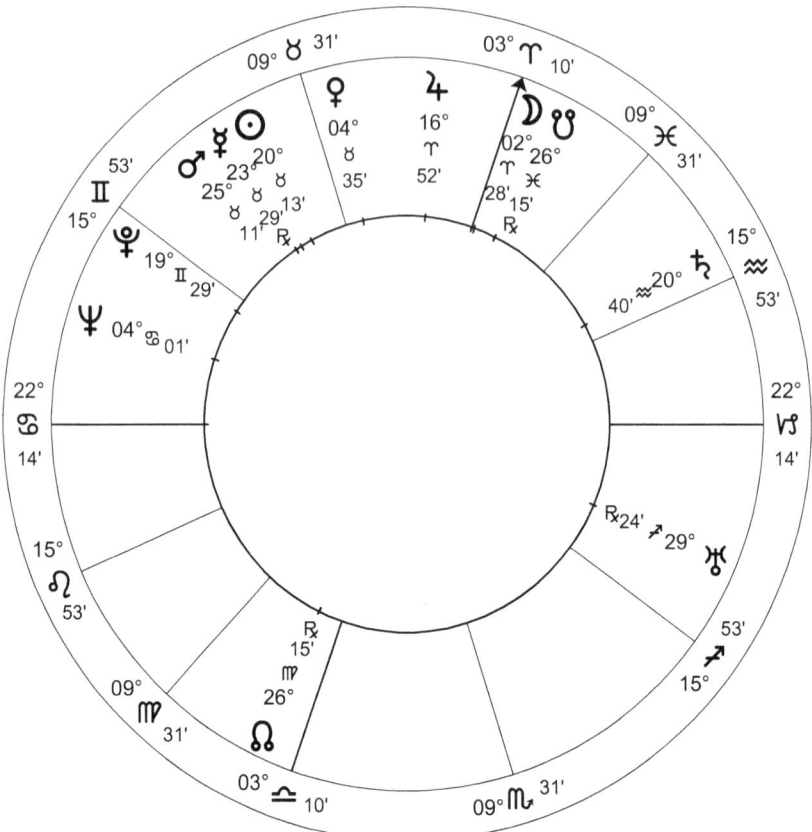

Salvador Dali. 11 May 1904, 09.45am, Figueres, Spain.[6]

she was not a noticeably maternal woman. But the maternal motif of an egg crops up a lot in Dali's work and in one surviving short art film the crazy artist and Gala can be seen cracking out of a giant egg on a seashore.

The Witch of Atlas
A contender for inclusion in our earlier chapter on 'The Witch' in the Old Moon section could have been Percy Bysshe Shelley's 1820 poem *The Witch of Atlas*. But this unearthly lady who lived 'within in a cavern, by a secret fountain' is far more reminiscent of a man's muse than an actual down-to-earth witch. This is the magical multiform romantic muse of a

---

6. Source: Birth certificate. *DataPlus UK* quotes Lois Rodden.

man born with the Moon in sensitive dual Pisces, (and in the 12th house by Placidus).

Shelley drowned in a boating accident aged 29 only two years after he wrote this poem and the circumstances of his death remain mysterious to this day. In the previous weeks around the time of his Saturn Return he confided to his wife Mary that he had met his doppelganger on the terrace. The figure mysteriously enquired 'How long do you mean to be content?'

Ghosts or human shadows are ruled by the moon, according to Paracelsus[7] and on the day of the poet's watery demise (8 July 1822) the Moon was in Pisces.

> 'A tapestry of fleece-like mist was strewn,
> Dyed in the beams of the ascending moon – and now she grew
> Pale as that moon, lost in the watery night –
> And now she wept, and now she laughed outright'
> [Percy Bysshe Shelley 1792-1822. From *The Witch of Atlas*]

'A woman colour'd ill', Shakespeare's Dark Lady

Shakespeare's sonnets numbered 127-152 [8] have come to be known as the Dark Lady sonnets as they appear to address an unnamed lady who is the mistress and muse of both the poet and his professed best friend. Whether or not she was an actual person has never been substantially proven either way. She is called the 'dark lady' from the description of her complexion, hair and eyes:

> 'her breasts are dun' [Sonnet 141],
> (her) 'hairs are raven black' [Sonnet 146],
> (she is) 'a woman colour'd ill' [Sonnet 153].
> 'Then will I swear beauty herself is black' [Sonnet 145].
> 'For I have sworn thee fair and thought thee bright
> Who art as black as hell, as dark as night' [Sonnet 138]

From these last and certain other lines some have argued that the dark lady carries a sexual disease. The poet certainly suffers for her favour although this could be a suffering in heart and mind at her muse-like fickleness. One sonnet in the sequence is quoted below in full as it has several lunar themes:

---

7. Quoted by Manly P. Hall in *Astrological Keywords*, The Philosophical Research Society Inc., Los Angeles, 1958.
8. The sonnets are numbered here according to the sequence suggested in Sir Denys Bray's *Shakespeare's Sonnet Sequence*, 1938.

'Lo, as a careful housewife runs to catch
One of her feather'd creatures broke away,
Sets down her babe, and makes all swift dispatch
In pursuit of the thing she would have stay;
Whilst her neglected child holds her in chase,
Cries to catch her whose busy care is bent
To follow that which flies before her face,
Not prizing her poor infant's discontent;
So runn'st thou after that which flies from thee,
Whilst I, thy babe, chase thee afar behind;
But if thou catch thy hope, turn back to me,
And play the mother's part, kiss me, be kind:
So will I pray that thou mayst have thy 'Will',
If thou turn back and my loud crying still' [Sonnet 132]

Mermaids

'Troop home to silent grots and caves!
Troop home! and mimic as you go
The mournful winding of the waves
Which to their dark abysses flow!'
      [George Darley 1795-1846. *The Mermaidens' Vesper-Hymn*]

She is not a being of the sea. She is not a being of the land. She is a being of the shore where shifting boundaries breathe with the moon. The mermaid is a dangerous muse.

The shoreline of the ocean is a magical place where creeping sands mark the meeting of ideas conscious and unconscious, of human life and fish life, of warm-blooded flesh and cold silvery scales. The ever-shifting lunar shoreline is where eons ago oceanic life adapted to live on land and in dreams as in reality the shore is a borderland between elemental worlds. It is a nexus, where two ways or destinies link. Like dusk and dawn, when night and day are balanced, or equinox when summer and winter meet, it is a temporary doorway into a different land. But like other doorways the shore has one factor on whose rulership all will agree. The ebbing and flowing tide-line comes directly under the influence of the moon.

'…she was gone; gone down the tide;
And the long moonbeam on the hard wet sand
Lay like a jasper column half-uprear'd'
      [Walter Savage Landor 1775-1864. *The Sea-Nymph's Parting*]

A mermaid symbolises the union of land and sea because she can live in both elements. Yet, as she is only half a woman and half a fish, she is not completely at home in either. She sits on rocks, in air she can breathe, while her tail rests in waves. She is a portal-keeper to an otherworld that is not wholly the sea, but is very likely the world of the moon.

The mermaid of folklore is engrossed in her own reflection, combing her hair. Her mirror fashioned from a shell or stolen from the domain of people is a silvered moon-ruled object that allows her glimpses back into her parallel realm. If ever there was a borderland creature to mark the entrance to a parallel world, it is she.

The Mermaid of Zennor
From all the folk tales and legends of travellers being drawn to the siren's song and then lost in her unearthly world, there is one that is unique. The village of Zennor in Cornwall holds the legend of a mermaid who reversed the normal way of things. It was the mermaid who was drawn away from her element to try and live in the world of the human.

From her home on the shore the mermaid of Zennor took a liking to one of the handsome young men of the village, and with a cloak hiding her tail would come into town just to be near him. It was his voice singing in the choir that had first attracted her, drawing her to him in a turnabout of the time-honoured siren practice. Each Sunday she appeared in the church in which he sang, taking her pew at a discreet distance from the rest. The strikingly beautiful figure in a long silver dress held her songbook clumsily but gazed at her love in admiration and wonder. Towards the end of the service, when everyone was bent in prayer, she would slip away and not be seen until seven days had passed.

She became the talking point of the village. Dark silvery women who appear from nowhere and vanish to nowhere were not everyday happenings, even in a fishing community weaned on sailors' tales and smugglers' yarns. Neither did the young man in the choir remain long unaware of how her eyes gazed yearningly at him, and he soon lost interest in all else and lived only for the Sabbath when he could sing out his feelings to her. Halfway through the Lord's Prayer he knew she was still looking. After 'Lead Us Not Into Temptation' he would peep up and see her face still smiling above the bobbing tide of bonnets.

Then one day without warning the young man had gone. There was no note at his home, no explanation. And the following Sunday the silvery woman was absent from church, never to return. Then the people knew she was a mermaid and had stolen her lover away.

And they carved her likeness in the choir stall, still there for all to see, and for many generations they heard her singing in Vear cove along with her husband and children.[9]

Rising Moons: The Ship's Figurehead

The elaborately carved and painted figureheads on the prow of sailing ships embodied the spirit of the vessel they fronted. Usually a human female figure (less commonly an heraldic beast), they were an inspiration to the crew and it was genuinely believed that the figurehead protected the ship, which was also a 'she'. To sail on a ship without a figurehead was thought unlucky, a superstition rooted in the belief that if the ship went down there was no overseeing spirit to guide the dead souls to the hereafter, or to the moon – the abode of souls.

Other sailors' superstitions included the taboo against changing the name of a ship. The figurehead and the name are closely connected, often painted and displayed together at the cutting edge of the vessel. Astrologically both a name and a figurehead would be represented by the ascendant in a horoscope.

The prow of an astrological chart is the ascendant and a rising Moon is an obvious figurehead. However you can turn any chart to put the Moon on the ascendant and then see in which areas of life (houses) we can express that inspiration through the positions of the other planets.

The Silver Lady

Following the custom of figureheads on ships, pioneering motorcar owners in the early twentieth century fixed personal mascots on to the front of their cars and statuettes of women were the most popular. The infant Rolls-Royce company observed the custom and decided to manufacture and provide their own special mascot on the front of every car they produced. This was the famous 'Silver Lady'.

We don't know the time that Charles Rolls and Henry Royce shook hands at the Midland Hotel, Manchester on 4 May 1904 but from that date their association was secured. The partnership of Rolls-Royce was born on that day, leading to the first Rolls-Royce motor car being shown publicly in

---

9. Details of this and other Cornish mermaid stories from *Mysteries of the Cornish Coast* by Ian Addicoat and Geoff Buswell, Halsgrove, Tiverton, 2003

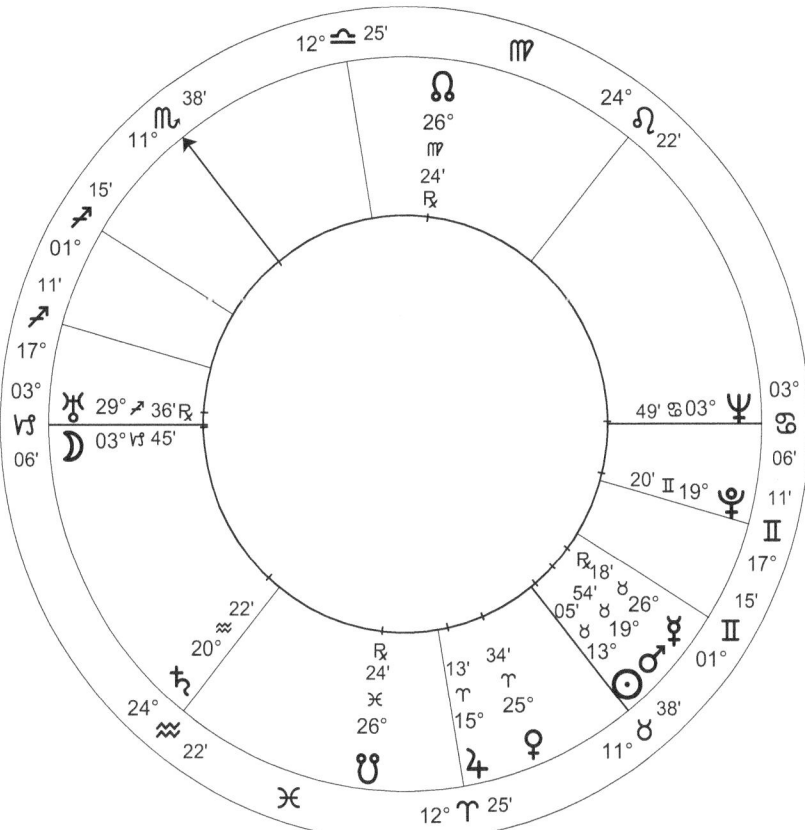

Rolls-Royce. 4 May 1904, 00.00 hours, Manchester, England

December of that year and eventually the founding of Rolls-Royce Limited on 15 March 1906. As an historical date 4 May 1904 has entered the record books in connection with the soon-to-be famous Rolls-Royce automobile company, and our chart is set for zero hours on this day.

Although this is a zero hours chart we cannot miss the exact rising moon. The automobile company first gained respect and renown with a model named the Silver Ghost, and throughout the years many other models retained the 'silver' epithet: Silver Wraith, Silver Dawn, Silver Cloud, Silver Shadow, Silver Spirit etc. The Silver Lady figurehead, known initially as the Spirit of Ecstasy, was positioned above the radiator of the car, appropriately where the water is held. The rising Moon in Capricorn suggests respect and longevity and a tasteful female figurehead/mascot that quickly became a status symbol. As the ascendant is both the figurehead and the name, the 'R' in Rolls-Royce fits well too as 'R' is the 18th letter of the alphabet. (Rather like the 18th Tarot card being The Moon).

The Moon on this chart is opposite Neptune, which is itself conjunct the descendant. One meaning of this could be the disappearance of a partner. After founding the automobile company with Henry Royce, Charles Rolls died in an aircraft accident little more than five years later aged 32 (12 July 1910). Even more relevant to the moon symbolism was the watery disappearance of the real Silver Lady.

The Silver Lady figurehead had been modelled on a flesh-and-blood woman called Eleanor Thornton, an actress and artists' model who was the secretary and mistress of the second Lord Montagu of Beaulieu.[10] In 1915 during the time of the First World War, and at her age of 35, Miss Thornton was on a passage to India when an enemy submarine torpedoed her ship and she was drowned at sea. It is oddly fateful that the First World War was the last time that figureheads were used on ships.

Rolls-Royces are still sold with a Silver Lady on the front despite obeisance to modern safety considerations and the forces of aerodynamics. The lady is scaled down now and retracts from sight if necessary at the press of a button inside the car, but she is still the spirit of the vessel, a protective grandeur that most other cars lack. Many individual car owners make up for this loss by dangling a homely mascot of some kind inside their front windscreen or in their back window. Lorry drivers sometimes attach a mascot to the front radiator grille. Cuddly toys are common; little animals or 'furry' dice. Childlike moon things make an unconscious connection with security and safety. The best mascot to hang in your own car would be one that somehow represented your Moon sign.

The rising Moon on the Rolls-Royce chart was technically a waning one, though exactly trining the Sun in luxury and materially minded Taurus. (Mars and Mercury in Taurus too). The Moon's rising status gave it a leading edge, true of any moon phase if the Moon is on the ascendant. The *phase* of the moon takes second place to the general strength of a rising moon. For example people born with the Moon rising will often have a round moon face despite its actual phase on their chart. The old rhyme 'Monday's child is fair of face' may more likely have been rendered originally as 'Monday's child is *full* of face', reflecting the Monday-Moon connection.

---

10. Beaulieu in the New Forest now houses the UK's National Motor Museum.

Ill Met by Moonlight: The Moon and Fairy Queens
The quarrel between Oberon and Titania, the king and queen of the faerie realm in *A Midsummer Night's Dream* is the underlying cause of the chaos and confusion that will soon abound in the human realm to those wandering in that same wood on the same night.

'Ill met by moonlight' are the opening words of Oberon to Titania when they meet accidently, still rancorous and resentful towards each other. And it is a waning moon, four days before New; the first lines of the play tell us this clearly as the betrothed lovers Theseus and Hippolyta are counting the days before their marriage. 'How slow this old moon wanes,' laments Theseus to his lady. She tells him they only have to wait four more days and nights before their wedding at the new moon.

In a deep sense Oberon, the fairy king, represents the sun and Titania, the fairy queen represents the moon. Their disharmonious state at this dark moon time is reflected in the male/female balance going awry in the outer world. The fairy queen, the moon, is cold towards the sun, and he is cold to her. The sun feels shut out; the moon is in no mood for reconciliation. Then to add to the derangement the queen falls in love with an ass…

All is resolved by the new moon however. The sun and the moon and all male/female counterparts come together for the wedding feast and a multiple marriage union.

William Lilly and the Queen of the Fairies
"*Regina Pigmeorum, Veni!*" (Come, Queen of the Fairies), was the magical evocation chanted by the 17th century astrologer William Lilly (natal ascendant Pisces, Sun conjunct Venus in Taurus, Moon in Capricorn) when summoning the feminine force to appear in Windsor Forest, according to W.B. Yeats in a footnote to his *Celtic Twilight* collection.[11]

Lilly's own autobiography mentions this as a call that Ellen Evans, the daughter of his tutor, would make into her crystal when wishing to converse with angelic creatures, but he advised us that 'it is not for everyone' and he doesn't make it absolutely clear whether he used it himself, successfully or otherwise.[12] He may be trying to distant himself in these recollections in his

---

11. W.B. Yeats *The Celtic Twilight*, Colin Smythe Limited, Buckinghamshire, 1981. (Originally published 1893. The footnote on Lilly dated to the 1924 edition).
12. *William Lily's History of His Life and Times, from the Year 1602 to 1681*, the autobiography edited by Elias Ashmole. From the edition published in London 1715, available online from Project Gutenberg.

old age from practices that others might frown on but he does mention an encounter in his own Hurst wood when a friend called up the Fairy Queen and was answered by an increasingly great wind and the appearance of the Lady in her 'most illustrious glory'. The supernatural experience was too great for the friend to bear and again it is unclear in the passage whether Lilly himself was present at the scene or was simply told about it afterwards. He does not say at which moon phase this occurred or which moon phase might be regarded as best for such an undertaking.

Gloriana, the Faerie Queene
Queen Elizabeth I (1533-1603), some of whose own heart-felt poetry still survives, was herself an inspiration to many poets in her day. She was allegorised as 'Gloriana' in Edmund Spenser's epic poem *The Faerie Queene* (1590) and specifically as Cynthia the moon goddess in Walter Raleigh's *The Ocean's Love to Cynthia* (late 16th century unfinished). As a regal virgin queen who loved to adorn herself with pearls she personified the moon to her people and few monarchs have ever inspired such genuine poetic outpourings from their subjects as Elizabeth.

Her acceptance of astrology is well-known. The Moon is exalted in Taurus on her birth chart and trines the ascendant for the recorded birth time of 'three o'clocke in the afternoon', Sunday 7 September 1533 at Greenwich.[13] Although Elizabeth's mother, Anne Boleyn, was executed by her father, Henry VIII, a strong sense of family loyalty is present with the Moon's fourth house position. The Moon also rules the seventh house, which contains Saturn.

The Queen never married, maintaining she was married to her people, and the natural public popularity of a strong moon is evident.

---

13. The 3pm time of birth is quoted by Wanda Sellar in 'The Nativity of Queen Elizabeth I', *The Traditional Astrologer*, Winter 1994, but there are various other charts in existence quoting slightly different times. All however give a birth time around 3pm with the exalted Moon usually in the 4th.

The Owl and the Pussy-Cat
Edward Lear's enduring moon poem

Was the Pussy-Cat the Owl's muse?
Let's spot the lunar references.

> 'The Owl and the Pussy-cat went to sea
>   In a beautiful pea green boat,
> They took some honey, and plenty of money,
>   Wrapped up in a five pound note.

> The Owl looked up to the stars above,
>   And sang to a small guitar.
> 'O lovely Pussy! O Pussy my love,
>   What a beautiful Pussy you are,
>   You are,
>   You are!
> What a beautiful Pussy you are!'

> Pussy said to the Owl, 'You elegant fowl!
>   How charmingly sweet you sing!
> O let us be married! too long we have tarried:
>   But what shall we do for a ring?'

They sailed away for a year and a day,
  To the land where the Bong-tree grows
And there in a wood a Piggy-wig stood
  With a ring at the end of his nose,
    His nose,
    His nose,
With a ring at the end of his nose.

'Dear pig, are you willing to sell for one shilling
  You ring?' Said the Piggy, 'I will'.
So they took it away and were married next day
  By the Turkey who lives on the hill.
They dined on mince, and slices of quince,
  Which they ate with a runcible spoon;
And hand in hand, on the edge of the sand,
  They danced by the light of the moon,
    The moon,
    The moon,
They danced by the light of the moon.'

> [Edward Lear 1812-1888. *The Owl and the Pussy-cat*
> first published 1871]

Owls and pussycats are actually both creatures associated with the moon, specifically the Black Moon (see Cats in Black Moon section). And the poem's first line is triply lunar:

'The Owl and the Pussy-cat went to sea…'
(Owl, Pussy-cat, Sea)

'In a beautiful pea green boat'

A boat is certainly lunar, although we can't be so sure about the 'pea green' part. Edward Lear used the phrase 'pea green' often in his writing, and while many green vegetables are said to be moon-ruled, like cabbages, lettuces, sprouts… peas are slightly different because they are climbers and have edible fruit/seed. Culpeper doesn't list 'peas' in his *Herbal* but al-Biruni gave them to Mercury.[14] Anyway the sight of peas in a pod is quite womb-like and Edward Lear's Moon was in Gemini so it ties in vaguely with al-Biruni's Mercury suggestion.

---

14. Dr J Lee Lehman *The Book of Rulerships*, Schiffer, PA.

In essence the Owl and the Pussy-cat sail away by night to an exotic land, a dreamland perhaps – an inner lunar landscape – and in this land where the Bong-tree grows they purchase a ring from a pig.

A silver ring would certainly be ruled by the Moon, as is a pig according to Ramesey in the17th century,[15] but we're not told if the ring is silver or gold. We are told that the pig was willing to sell his ring for a shilling, and a shilling was a silver coin, so that seems good enough. On the way to this magical land the Owl is overcome with romance and serenades the Pussy-cat under the stars to the accompaniment of a guitar.

It doesn't take much imagination to classify an acoustic guitar as a moon instrument either. Given that all musical instruments are affiliated to Venus because they produce harmonious sounds and soothe the savage breast, there remains a certain difference between the gentle lute or guitar and a bulkier object like a trombone or a an xylophone. A guitar is hollow like a container and its body is shaped like a woman.

The lovers are then married by a turkey. And this is unfortunate for us as turkeys don't figure in medieval lists of correspondences so we can't be absolutely sure what planet rules them. As a domestic fowl their nearest equivalent might be the goose, which is certainly moon-ruled. In the Pussy-cat's eyes the Owl himself was an 'elegant fowl'.

Lastly after a wedding feast eaten with the oddest of spoons (silver, concave, moon-ruled) they dance on the shore hand-in-hand by the light of the moon – which must surely be a full moon to light up their dance. And so important is the Moon that it is repeated several times at the end.

The loony Moon of Edward Lear was conjunct Mars in Gemini, opposite Neptune and square Pluto, all part of a mutable T.square. The force of the Mars-Moon in Gemini conjunction explains his need to play around with words, noticeable in the limericks for which in his time he was best known.

He was however an accomplished and serious artist and at one point gave lessons to Queen Victoria. His rising Saturn in Capricorn opposite a Venus-Jupiter conjunction in Cancer provided the discipline to draw and paint beauty with deliberation; his botanical studies are beautifully rendered and highly sought after. The rising Saturn also caused him to fear the outer world from an early age as he suffered from epilepsy and was embarrassed to be seen having a fit.

---

15. ibid

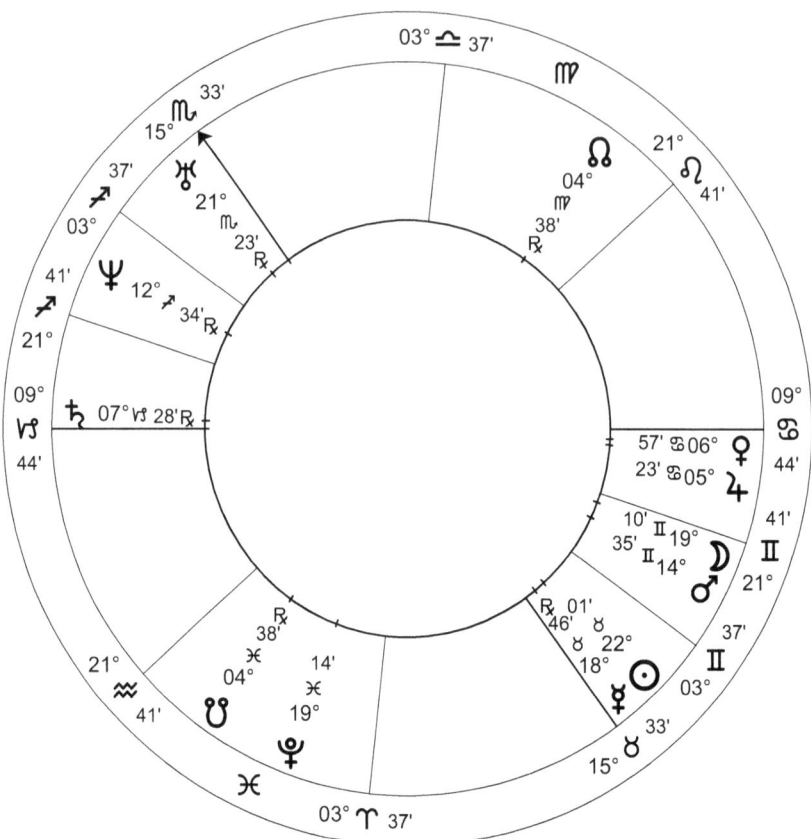

Edward Lear. 12 May 1812, 11.30pm, London.[16]

But Scorpio was on his midheaven and its rulers old or new were part of the lunar T.Square, not part of the Saturn-Venus opposition. His greatest fame would come from his nonsense verses. Not surprisingly Uranus was in his tenth house.

The Owl and the Pussy-cat begins with a voyage into a new phase of life. When the owl looks up to the sky he sees only stars, not yet the moon. If it's a new moon it would fit the general moon phase on Edward Lear's chart which is also new.

---

16. Source: Astrological Association database. Original source unknown.

Vintage Moon Fiction

Following are a few short summaries of well-known novels that might be seen to demonstrate different properties of the moon. The main criterion, other than their demonstrable lunar streaks, is that in titles at least these all come loosely under the category of 'household names', which is a good descriptive phrase for a popular aspect of the moon.

The moon as house and woman:
Rebecca by Daphne du Maurier. (Originally published in 1938)

'Last night I dreamt I went to Manderley again…' One of the most famous opening lines in modern English literature greets us as we open Daphne du Maurier's most celebrated novel.

There is a haunting atmosphere right from the start as we continue into several pages describing a dream visit to the now derelict house and gardens of Manderley. Somehow, although we don't know what emotional trials have taken and will take place in this Cornish estate with its lawns that stretch to the sea, nor what vital part the sea will play, a melancholy feeling seeps from the pages even in the descriptions of woods and flowers. An atmosphere of foreboding and loss pervades before we have fully entered the story.

And it is a story told in retrospect by a young heroine whose name we never learn. The reactions of others suggest hers is an unusual name, though it is overshadowed almost from the first by 'Rebecca', a woman she will never meet – the first wife of the man she marries. Subtly the lack of her own name suggests that people like her employer and then her husband are curt or condescending towards her; and this bolsters our impression of her naivety and lack of confidence in an adult world. She is a young New Moon.

After the opening dreamlike sequence we meet some important characters, not at Manderley but far away on the French Riviera. Each of them is moon-like in different ways. The narrator herself, a shy young lady's companion, by her own admission unworldly and awkward. Her employer Mrs Van Hopper, a silly thick-skinned gossip whom everyone avoids. The mysterious Mr de Winter, the owner of Manderley, a gentleman above snobbery who treats our young narrator as a person, though his preoccupation with some secret past gives him a very moody edge. Later we meet the real ogre of the piece, Mrs Danvers, the Manderley housekeeper, who strives to live in the past and make our heroine's life a misery.

Daphne du Maurier's new moon birthchart (she was born on the day after a new moon) has the Sun in the seventh house allowing her to identify and portray a theme of one who must live through others or who loses their identity to another or gets trapped in another's life, and *Rebecca* is a clear instance.[17] The heroine is nameless, forever a pale shadow to the lovely lady of the title. She (the heroine) must struggle against great odds to assert her own identity. And the novel makes the beautiful but suffocating house of Manderley almost another character in its own right. Du Maurier's Moon was in Gemini, continually adapting itself to others, while its 8th house position suggests secrets in a dark ancestral home.

The work might equally have been titled 'Manderley'; it's mentioned in the first line (where 'Rebecca' isn't) and it's described in detail at the story's beginning. It's the setting for the bulk of the novel, and it will presumably 'die' in the end as we have met its ghost in the opening pages. The house (the moon) overshadows the ego of the heroine in the same way that Rebecca overshadows the ego of the heroine. The two are intertwined; house and woman. In that sense only is this 'a woman's story'. There must be a million pocket-romance fantasies where the virgin orphan marries the older landowner and becomes the Mistress of the House but although this present story is narrated by a woman and most of the main players bar Max de Winter are female, it is on a different plane. It's *Cinderella* with a darker verge. Du Maurier was never a straightforward romantic novelist and she railed against that false reputation. Interestingly male figures become more noticeable in the latter part of the book at the same time as the heroine herself begins to take an assertive role. The moon is both fuller and on the wane we might say.

I was not surprised to learn that Daphne du Maurier wrote most of *Rebecca* away from Cornwall. She was with her husband at a military posting in Egypt, uncomfortable and homesick, and explains why the opening of the story proper, set on the Mediterranean coast, does not paint an exceptionally glowing picture of those sunny climes. The wistful description of the gentle seasons of her homeland in comparison bears the longing of one for whom distance was lending enchantment to the view. Sally Beauman, an expert on this novel who has written a modern sequel called *Rebecca's Tale*, calls *Rebecca* 'a deeply subversive work'. Like the moon it has hints and depths

---

17. This theme is pronounced in many of Daphne du Maurier's novels and short stories: *Don't Look Now, The Scapegoat, The Rendezvous, A Border-Line Case, Split Second* etc.

that may reveal different faces to different people. It is unquestionably a classic.
[Daphne du Maurier. 13 May 1907, 5.20pm, London, England.[18] New Moon with Sun in Taurus in 7th, Moon in Gemini in 8th]

The moon as patriotism and home country
***The Moon is Down* by John Steinbeck. (Originally published in 1942)**
Steinbeck is a giant name in twentieth century literature and his most famous works speak about struggling workers and families trying to make a home on the land (*Of Mice and Men, The Grapes of Wrath*). The social messages underlying could be as powerful as the plot and characters themselves and made Steinbeck a controversial author in his day. This novel *The Moon is Down* about an unspecified country invaded by another was written to help the American war effort in 1942 (he first wrote it in 1941 just before the attack on Pearl Harbour) and was deliberately meant to give out a message. Through ordinary small town characters the story shows in a light and readable form how a conquered people are never really conquered, how discipline and superior numbers can win the outer battle but how it is individuals who win the inner war.

Steinbeck, based in America, set the fictitious tale in a small mid-western town in his native land, and submitted it as a morale-boosting parable. But it was officially rejected for publication by the Foreign Information Service because of the unacceptable idea of America being under the yoke of an invader. In a later article called 'Reflections on a Lunar Eclipse' Steinbeck described how he then sought information from political refugees who had fled to the US from occupied Europe and rewrote the book in a different setting. It was now a European country invaded by a larger and aggressive neighbour, although neither nation was named and it was still his intention that the basic characters could apply to any country, even his own. Acceptable now to the American authorities it was published in 1942 but to his dismay immediately received harsh and violent criticism from some quarters for portraying the invaders (obviously the Nazis) in too human a light. In other words not all the individuals in his occupying army were manic inhuman psychopaths. Yet Steinbeck had produced a piece of propaganda far more powerful and long ranging than anyone realised

---

18. Source: AA *Transit* magazine July 1996 quotes birth time from the biography Daphne du Maurier by Margaret Forster, 1993.

and the full effect of his novel would not be known until several years had passed.

In Britain this book was available openly but in almost every country in occupied Europe it was secretly translated into the native languages by local resistance movements. Separately in Norway, Denmark, Holland, France... pirate copies were reproduced in underground cellars and distributed secretly to the people in defiance of the enemy. The novel came to be regarded as so dangerous by the Axis powers that in Italy at one point ownership of a copy carried an instant death sentence. Every occupied European country thought that Steinbeck had written it with them specifically in mind as it described the situation of their daily lives so well, with local collaborators often regarded as worse than the enemy itself. Unlike the usual patronising heavy-handed wartime propaganda this was morale boosting because it was individually empowering and more real.

Receiving honours in Norway after the war Steinbeck was asked how he had understood their plight and their spirit so accurately from his cultural and literal distance. He replied that he put himself in their place and thought what he would do. And that may go some way in explaining why his work in general is regarded so highly. It is in this category of the human condition that his greatness lies, rather than just as an accomplished poet or artist. Few can triumph in both disciplines, except someone like Shakespeare I suppose. But *The Moon is Down*, which actually is a quote from Shakespeare's *Macbeth*, is a deceptively powerful and moving piece of work about the unshakeable strength of one's home country and cultural roots, presented in a subtle and simple way.

[John Steinbeck. 27 February 1902, 3pm, Salinas, California, USA.[19] Moon in 4th conjunct Node trine Sun]

The moon as dark closeted passion:
***The Picture of Dorian Gray* by Oscar Wilde. (Originally published in 1891)**
Everyone knows the basic plot here: how a man's portrait grows old as the man continues to remain ever young. But it seems incredible, reading it now, that no one realised at the time of its first publication – five years before Wilde was tried in court for homosexual acts – that the author was gay.

---

19. Source: Lois Rodden, *American Book of Charts*, 'C' rating.

The story begins at an artist's studio opening onto a summer garden where, amongst the sensual luxury (some might say decadence) of flowers, Persian divans and opium cigarettes, two young men discuss a third beautiful acquaintance named Dorian Gray, whom one of them is painting. The physical descriptions of male beauty plus the obvious jealousy between the two men to enjoy the company of Gray are difficult to fully understand other than that the three men are homosexual. One of these young men, Lord Henry Wooton, who is neither the artist nor the sitter, speaks in that witty worldly Wildean way that leads us to assume that this may be Wilde himself in an ideal world he would like to be. This character, like Wilde, is married yet ever pontificating on the uselessness of marriage and spends more time with his male companions than he does with his wife. And it is he who talks the young Dorian Gray into letting loose his soul and freeing his deepest self from the constraints and inhibitions of convention – in today's parlance we might say to 'come out', but it is clearly a little more than just this – and it happens while the portrait is being painted by an artist who is romantically obsessed (read: 'in love') with him too. The resultant moment of highly charged emotion is captured forever on canvas. The painting is a masterpiece and the whole novel explores the complicated interplay of art and life. (It was Oscar Wilde who first coined the phrase 'life imitates art').

Within the languid gentle prose we find a reference to the moon as 'that silver shell'. The great feminine force is likened to a shell – surely that doesn't imply an empty thing? One can argue that the Moon is always strongly identifiable on the charts of gay men, in the sense that they often live out their Moons more noticeably than heterosexual men. Oscar Wilde's chart had a Leo Moon exactly square Uranus.

Dorian Gray is a narcissist who fears growing old but finds that every bad act or moral sin on his part changes his portrait accordingly. He locks the portrait in an unused room so that only he knows its secret, finding he can free himself from remorse or guilt or 'sin' as the portrait obligingly takes on the suffering. In a way it is a new twist on *Jekyll and Hyde*, written six years earlier; a man split into a light half and a dark half, except that here all impulses reside in the one man. His painting cannot stalk the streets.

Of the three men only the artist is a character with whom we can properly sympathise, and it is likely that Wilde saw parts of himself in all three of these protagonists. The terrible burden of having to hide a guilty secret weighs heavily behind the entire plot, and the many lesser 'sins', which are often hinted but not spelt out, are easy to read as those that at that time never spoke their name.

Dorian Gray's birthday is mentioned – the tenth of November. The secretive man who explores the depths of human passions and the bounds of life and death is a Scorpio.
[Oscar Wilde. 16 October 1854, 3am, Dublin, Ireland.[20] 12th house Moon in Leo square Uranus]

The moon as the power of the sea:
Moonfleet **by J. Meade Falkner. (Originally published in 1898)**
First published in 1898 and telling a tale set over a century before – 1757 in the coastal Dorset village of Moonfleet – 15-year-old John Trenchard, orphaned and living with his aunt, stumbles on and literally into, the dark pervasive ancestry of the Mohunes, once the lords of the land. A great storm and flood, the worst since the reign of Queen Anne, is followed by strange bumpings under the old church during Sunday service. The rector explains it as coffins afloat in the crypt, but the local folk believe it to be the unquiet spirit of the disgraced Colonel John Mohune still restless in the family vault. It was he who had been the Parliamentary Governor of Carisbrooke Castle on the Isle of Wight when King Charles I was held in captivity, and he (Mohune) who had cheated the king out of a priceless diamond. Our young hero makes a midnight visit to the church, descends a passage into the fearful vault and discovers more hidden skulduggery than ghosts and bones. Smuggling, wrecking and the might of the sea carry this powerful story to its satisfying conclusion. As it says in the Latin inscription carved in the local Inn's backgammon-board: 'As in life, so in a game of hazard, skill will make something of the worst of throws' – and greed and a cursed stone will do the opposite it seems. It's hard to believe this book was written well over a hundred years ago; the pace is fast, the language easy, the 18th century turns of phrase vital and poetic. It rivals *Treasure Island* as a timeless sea-bordered adventure.

The moon as old wives' tales and loony families:
Cold Comfort Farm **by Stella Gibbons. (Originally published in 1932)**
On the death of her parents a typical Bright Young Thing of the early 1930s goes to stay with a Bright Young London friend, a Mrs Smiling who has a hobby of collecting bras (or brassières as then known), starting our moon associations right away. But after a while our heroine feels it would be more productive to inflict herself on country relatives she'd never met

---

20. Source: From baptismal certificate.

so consequently travels to the mad family of Starkadders overseen by the ancient matriarch of Cold Comfort Farm.

Although some of the parody of rural novels of the time may now be lost on us, *Cold Comfort Farm* has remained unfailingly popular for the past eight decades. It's full of original funny touches. The pretend Foreword explains how after being a journalist and learning to get to the point in short sentences the author has now had to try her hand at a more serious literary art and has accordingly marked with asterisks paragraphs throughout the book she feels are eminently noteworthy and that reviewers might like to home in on to make their jobs easier. And such overblown paragraphs do crop up now and again as we're reading the story, making us smile not only at the florid description but at the artfulness of it all.

Cold Comfort Farm was Stella Gibbons' first book, written when she was in her late twenties, and popularly – and erroneously – thought to be her one and only. In fact she wrote a further twenty-three, but few remained long in print. It's an odd situation for an author when their very first baby is not only an instant hit but remains a popular bestseller and no matter how much their subsequent writing grows it can never measure up to that first creative outburst. And it's true that just about everyone seems to love *Cold Comfort Farm*, especially women, happily quoting the daft names of its cows and characters, its saucily written turns of phrase, its breezy slaying of old wives' tales, and its 'something nasty in the woodshed'. The heroine, who's the same person at the start as she is at the end, has no conflicting inner defects, no skeletons in the closet; she swans through the whole thing sorting out the emotional stick-in-the-muds at the farm with little fuss, then flies off with her boyfriend and leaves everyone to it.

Many people read it regularly as a kind of tonic.
[Stella Gibbons. 5 January 1902, London, England, time not known. Waning Moon in Scorpio or Sagittarius]

The moon as ghostly dog and family curse:
***The Hound of the Baskervilles* by Sir Arthur Conan Doyle. (Originally published in 1902)**
Sun Gemini Conan Doyle wrote this, one of his few full-length Sherlock Holmes adventures and probably the most celebrated, in 1902 before the great days of the whodunnit, and it is surprising to see that not everything is left to the last page. The mystery of the ghostly ancestral Hound from Hell who howls on the moor and haunts the Baskerville family is held back until

the climax and we're kept guessing at most points through the narrative, but the actual murderer has been deduced for us before the finale on the bleak and eerie sweeps of Dartmoor. Modern adaptations have usually given a happier end to the love story between the young Sir Henry Baskerville and his neighbour 'Miss' Stapleton; a theme that is left hanging in Conan Doyle's original.

There are some great exchanges between Holmes and Watson and had you never read any of the Sherlock Holmes short stories you would find everything here. Holmes uses much intuition in his deductions and a key to this mystery hangs, appropriately to our lunar theme, among ancestral paintings lining the hall. Vague references to supernatural undertones are met early on, but perhaps strangely for an author who would spend the rest of his life promoting spiritualism and unseen worlds, Conan Doyle keeps a firm hand on the tiller and plays out all the Sherlock Holmes stories with straightforward explanations. Sherlock Holmes has some weird cases but they all turn out to have a rational vindication. The 'severely practical mind' of which Watson sometimes accuses his cerebral partner is a highly inspirational one that allows space for any possibilities, whether real or unreal. Holmes never pooh-poohs the supernatural, which is perhaps why this story has long remained such a popular favourite. It is Conan Doyle on top form.

[Arthur Conan Doyle. 22 May 1859, 4.55am, Edinburgh, Scotland.[21] Moon at zero Aquarius trine Sun at zero Gemini]

The moon as habit and nostalgia:
Goodbye Mr Chips **by James Hilton. (Originally published in 1934)**
The author James Hilton was in his early thirties when he delivered this well-loved story about an ageing schoolmaster called Mr Chips. It's not a long book and Hilton is said to have written it in four days. But it's been adapted for stage, television, filmed twice, and never been out of print. Set, mainly, in an English boarding school it looks back over the years of Mr Chipping's life, principally his school life in the Victorian/Edwardian era, against the passing memories of history. It doesn't sound much admittedly. Interesting maybe in 1934 to view the past through this lens and observe how the old habits and values were changed forever after the First World War; but what has made *Goodbye Mr Chips* such a perennial favourite?

---

21. Source: AA Database has birth time of 05.07.52am

Partly and very largely it is the strength of the main character, a kind dependable man, moonlike in a homely way, loved by his pupils past and present. He is fond of inviting people to tea around the hearth. It's a good antidote to the general impression of boarding school life being all Tom Brown's Schooldays. But also it's a piece of work that flowed from the author almost as fast as he could type or write it, and therefore has that special quality of a product from the heart, from the mosaic of mind and memory, rather than a project carefully and logically planned. From a competent author this kind of book is always easy to read. In addition *Goodbye Mr Chips* has something very touching, trustworthy and peaceful about it; reviewing a rather ordinary man's rather ordinary life with its turns of sadness outweighed by its joys. The joys in little things, unchanging things, and the spreading of good humour to others all combine to evoke a sense of security that we can imagine has seen many a reader through their own dark patches. It's nostalgic to read even though we no longer resonate with the era of its setting.

[James Hilton. 9 September 1900, 02.10am, Leigh, England.[22] Full moon in Virgo/Pisces]

The moon as a child's world of security and enchantment:
***The Wind in the Willows* by Kenneth Grahame. (Originally published in 1908)**
The willow tree is ruled by the moon and this tale of the little creatures who live under its weeping branches by a stream is one of the enduring standards of children's literature. Those good friends Rat, Mole, Badger and Toad always seem to be visiting each other's homes – warm snug little places, full of good food and good cheer on fine spring mornings or a snowy winter's night. Even Toad doing time in his prison cell for reckless driving is in a similar dark womb-like place, (from which he escapes dressed as a woman). The changing seasons produce their own long magical worlds to our little friends, just as they do to children. In the chapter titled 'The Piper at the Gates of Dawn', Rat and Mole row out on the river by moonlight to visit the island of the great nature god. As they land and fasten their boat in this 'silent, silver kingdom' the moon is described as 'she' and we sense that she watches protectively over all.

[Kenneth Grahame. 8 March 1859, 10.15am, Edinburgh, Scotland.[23] Sun in Pisces, Moon in Taurus – fantasy rooted in nature]

---

22. Source: AA Database
23. Source: Rodden from birth certificate.

The moon as a deadly nursery rhyme:
And Then There Were None **by Agatha Christie. (Originally published as *Ten Little Niggers* in 1939)**
Claimed as the best-selling thriller of all time, with over 100 million copies sold, it's obvious why the original racially unacceptable title was later changed to *Ten Little Indians*, and then renamed again *And Then There Were None*. But central to the entire plot is the deadly nursery rhyme of *The Ten Little Niggers*, or *Ten Little Indians*, each of whom dies in turn so that in the end 'there were none'.

In the story china models of these figures occupy a prominent place in the mysterious house to which the ten guests have been lured, each model enigmatically disappearing after a real murder has taken place. In its current wording for the 21st century the Indians have been changed to 'soldier boys', the models are soldiers, the nursery rhyme pinned up in each bedroom begins 'Ten little Soldier Boys', and the island itself is called Soldier Island. None of this feels right at all.

It need hardly be said that if this hadn't been one of Christie's best stories the political incorrectness may have led to it being dropped from anthologies and general reissue years ago. But it is not the story which is politically incorrect, it is the nursery rhyme, and the author didn't invent that, it was as familiar to everyone at one time as *Ring a-Ring a-Roses* or *Jack and Jill*. And you do wonder what other subtleties have been lost with the name changes. The name of the island is mentioned many times in the opening stages, overly so, as if the stage is being set up for something, and 'Soldier Island' just does not inspire the same sense of mystery and darkness. This modern tampering with the original text is annoying and renders some of the dialogue unintelligible. At one point two women guests are in a conversation that touches obliquely on African natives, when one of them becomes suddenly hysterical. Her fearful outburst can only refer to the fact that the china figures disappearing one by one are ornaments of little black boys, but with the rest of the text changed to 'soldier boys' and the models being soldiers rather than a disquieting set of antique ebony-skinned exotica, the panic it triggered has absolutely no meaning. (There are no black people in the story).

In any event this is a claustrophobic tale of people trapped in a house on an island, all of whom have secrets in their past. Filmed versions have either made the house seem creepy or relocated it to a foreign locale but the original written version comes as a surprise in that the house is newly built, brightly lit and ultra modern. There is much description of the house and its beautiful fittings and furnishings, deliberately at odds with the uneasy feeling of the place. The point is being made that the house is not to be regarded as the enemy; there is no extra person hiding behind secret panels or inside suits of armour so the murderer must be one of the guests. By deliberately rejecting the spooky house cliché the author succeeded in further disturbing her readers by introducing an unexpected feeling of unease into their own personal ambience of safe armchairs and electric lit homes.

The ending of the story is different to all the filmed versions and it is debatable which is the better. Hollywood required the lovers to survive; Agatha Christie had them all killed off so that 'then there were none'. I later found that it was Christie herself who gave the alternative ending of the surviving couple when she adapted her novel for the stage a couple of years after its first publication, and all subsequent film versions followed the play rather than the book. Her reason was not simply to attain a happy ending but to have a remaining character able to voice an explanation to the puzzle, to work out for us 'who done it'. As this was not an adventure of Hercule Poirot or Miss Marple or any other recurring detective there had been no typical summing-up scene at the end in which everything was explained. Suspects and victims were all dead so an alternative narrative device had to be found. Not entirely satisfactory in the book is a conversation between two police detectives trying to piece together what has happened on the island when the ten bodies are found. This leads to a last chapter, which is better, and is written as the contents of a note found in a bottle floating at sea, penned or typed before death by the houseguest who was the instigator of the murders. It's a bit long-winded for a note in a bottle, but has to be fully detailed from the reader's point of view to explain the murderer's motives and methods and to tie up all the loose ends. With the answer revealed inside a bottle at sea it subtly rounds off the moon themes of nursery rhymes and an isolated paranoiac house.

[Agatha Christie. 15 September 1890, 4am, Torquay, England.[24] New moon, Sun in Virgo, Moon in Libra (and ascendant Virgo) – Mercurial puzzles with the need for justice and fair play.]

---

24. Source: Astrological Association chart database.

The Moon as femme noir:
Double Indemnity by James M. Cain. (Originally published in 1936)

Single man meets married woman. Single man and married woman have passionate affair. Single man and married woman decide to murder married woman's husband to get him out the way and cash in on the insurance money.

This is the underlying plot in each of James M. Cain's celebrated Depression era crime novels *The Postman Always Rings Twice* (1934) and *Double Indemnity* (1936). They are similar in design – a first-person narrative told in regretful flashback by a single man. But *Double Indemnity* is slightly fatter than *The Postman Rings Twice*; a little more wordy, not so sparse, better heeled. Frank, the fictional narrator of the first book, was an out-of-work drifter while Cora, his female nemesis, waited tables in her husband's struggling roadside diner. They were young, simple, unsophisticated people drawn together in crime. In *Double Indemnity* everything moves up a notch. The narrator, Walter, is an insurance salesman, fifteen years in the business; the house he calls on is well furnished, its absent owner a man in the oil-drilling world. The married woman is in her early thirties with a stepdaughter of nineteen. It's a more grown-up scenario. This couple plan and fulfil the perfect murder, in contrast to Frank and Cora's improvised affair. This woman, Phyllis, has a steelier side – a more frightening dark moon side. Phyllis is a blacker widow than even we might at first perceive.

The book explores how love turns to hate when fear enters the picture, and now a vast difference opens between the couples in the two stories. In *Postman* Frank and Cora's emotional needle oscillated rapidly but they were basically lovers, in it together till the end. Walter and Phyllis are a different proposition. The sexual spark of their initial meeting triggers the opportunity for long-standing personal schemes. Walter intuitively knew she wanted to get rid of her husband before he (Walter) had ever arrived, and the heat of his lust drew him to suggest a perfect insurance scam, an idea that had occurred to him over the years but with never an intention of carrying out. Their mutual passion – and we are never entirely sure if Phyllis wasn't faking her side of it all along – quickly fades and other people enter the picture to play larger incidental parts, and therefore complicate the basic issues. Not that *Double Indemnity* has a deeply complicated plot; the beauty of it is that there *are* only a handful of main players and the perfect murder goes without a hitch. The tortuous landscape is within the minds of the people involved, their later interactions with each other, the secrets that emerge and the total unpredictability of fate.

And Fate may be what the book is really all about. Unpredictable fate. Cain (the author's name is eerily the same as the biblical Cain who in medieval times was seen in the markings of the moon in the sky carrying thorns on his back) exposed the dark weave of fate with a minimum of fancy prose. His plots were Greek tragedies relocated to a young straight-talking California. You can make your perfect plans but the gods have their own agendas. Walter Huff, the reliable law-abiding insurance man, steps into a world that operates under its own laws when he commits murder and sets unknown events in motion. Only on the last page does the author drop a hint of the true realm we have entered when he uncharacteristically, for such a bare-bones style, references poetic scholarship. Phyllis makes a last appearance and he says she looks like the entity that came aboard ship to shoot dice for souls in the *Rime of the Ancient Mariner*. We are in very strange waters now. Appropriately the last line of the book, presumably the last thing ever written by the story's fictional narrator, comprises just two words:

'The moon'.

[James M. Cain. 1 July 1892, Annapolis, Maryland, USA, time not known. Moon in Libra squaring Sun]

The moon as femme fatale:
***The French Lieutenant's Woman* by John Fowles. (Originally published 1969)**
The novel opens with a young man and his ladylove, Charles and Ernestina, promenading the Cobb at Lyme Regis in 1867; but from the first we are introduced to a third figure who stands windswept and dark robed staring endlessly out to sea – the French Lieutenant's Woman. Charles tries to speak to her but gets no response. If you strip away the buttoned-up Victorian moral overtones this is a classic *noir* novel where the man risks losing everything in his passion for a 'tainted' woman. Sarah, the so-called French lieutenant's woman, is infuriatingly enigmatic at the same time as being irresistibly magnetic to our hero and we never really know her. She advances and retreats, lures and rejects, tells half-lies and half-truths. She is the embodiment of that aspect of the moon and the difficulty in relating to it that hung over the educated male-dominated strata of the nineteenth century. Women were either ladylike ladies or common whores to a society that covered up the legs on a piano while brazenly displaying a pale-skinned nude on a painting above the hearth; provided she was an untouchable mythological figure. The novel is not written as a first person

narrative as this kind of heart confessional usually is, but with the author taking an omniscient stance as if conversing in period style but speaking to us from 100 years on, that is in 1967 when he is writing. We often step out of the Victorian story to contemplate comparisons made between the two centuries, and this runs from the personal level of dress sense to the collective aspirations and beliefs of society. Fascinated and amused as they were in the late 1960s by anything Victorian (Lord Kitchener, Sergeant Pepper etc), this shares the same exuberant irreverent spirit though with an accurate authenticity to Victorian period detail.

The mysterious Sarah becomes more known to us, and to Charles, and a triangle situation builds up between her and the betrothed couple. In a moonish way the author plays with us too, sliding in and out of the novel, inserting himself in a scene, and reminding us that he is creating all these players and they might do this and they might do that… So much so that he gives us a choice of endings to the plot. To our surprise the fiction even spills over into fact in the last chapters as we understand a further significance of Sarah's untamed red hair. The book is a *tour de force* and deservedly the most popular of John Fowles' works, which include *The Collector* and *The Magus*.

[John Fowles. 31 March 1926, Leigh-on-Sea, England, time not known. Scorpionic moon just past Full]

Lyrical obsession for a dark young moon:
***Lolita* by Vladimir Nabokov. (Originally published 1955)**
Great names in contemporary writing have praised this book to the hilt and there is no question, the prose is absolutely magnificent. It's a highly intelligent work with so many clever literary allusions and word plays you could spend months on this aspect alone. Hundreds of would-be novelists must have sighed, 'If only I could write like this' – but then again I'd imagine they'd also stipulate, 'but rather not on this subject'. For what kind of reception would this book arouse today were it being released for the first time? It's a story written by a middle-aged man (written from prison admittedly) who was, and still is, obsessed with a twelve-year-old girl. The man, well-educated and good-looking, goes by the unlikely name of Humbert Humbert, clearly an alias adopted for this account, while the object of his desire is called Dolores, or 'Lo' for short by her mother, or 'Dolly' by others, or 'Lolita' to the love-crazed Humbert. It is a comedy that sails close to the wind of pornography, allowable for its art. Nabokov introduces the word

'nymphet' to the English language in 1955; in fact he coins many words but none have stayed the distance like this one.

It's almost impossible to summarise *Lolita* without making it sound repulsive or distasteful, but somehow it isn't – quite. Perhaps today we are more aware of a paedophile's *modus operandi* than they were in the 1950s, and much of this fictional 'confession' seems uncomfortably true. Humbert for instance will go to extreme and devious lengths to get close to Lolita, which would be highly amusing in other circumstances if he were chasing someone his own age. And the way he likes to hang around school gates to ogle emerging young girls would perhaps have raised an astonished laugh in 1955 but is horribly familiar today. He even marries Lolita's mother to give him a legitimate excuse to lovingly fondle the daughter. And it gets worse. The pre-pubescent girl after whom the book is named (Dolores means sadness and distress and likens this young moon to a waning dark one) is the character we initially know least about. Her single mother finds her a wayward satellite, but she seems no more precocious than any other pre-teen to us. We are seeing her admittedly through the hungry fanatic eyes of one who worships every hair on her skin and every button on her clothes, and tells us about it in excruciatingly poetic detail. Only in Part Two of the book do we start to understand her personality a little more as lustful step-father takes newly acquired step-daughter on a permanent touring drive across America to stay in a succession of anonymous motels. But everything is still as recounted to us by this doomed jealous lyrical pervert Humbert.

In the last pages after the story has ended, Nabokov discusses some points about the writing of the novel. I was going to say he makes some excuses, but he doesn't really. He rightly defends an artist's freedom. The novel was initially turned down by a succession of publishers, hardly surprisingly, eventually printed in Paris and then lauded by literary critics everywhere. Of everything he wrote, and he wrote much in Russian and English, *Lolita* remains his most famous.
[Vladimir Nabokov. 4 May 1899, 00.59am, Petersburg, Russia.[25] Waning Moon square Uranus]

---

25. Source: AA Database.

The moon as reflection and imitation:
Invasion of the Body Snatchers by Jack Finney. (Originally published in 1955)

Destiny decreed that this little-known 1950s' science fiction novel and the modest black and white film that followed would be regarded in later decades as one of the most important in the genre. This is partly for its unintentional connection with the 'Reds-under-the-bed' phobia in the era of its writing, but in their time both film and book were just one more in a rich vein of post-war sci-fi dramas.

Invasion of the Body Snatchers did not truly become a household name until fifteen to twenty years later but the original book is a spooky tale in an ordinary small-town-America setting. It's written in the first person through the eyes of a young doctor suddenly swamped by local folks reporting strange things about their closest kin – they look the same, they sound the same but they are not the same. The story is now so well known that we already understand what's going on – aliens have landed and are infiltrating Earth families by taking over human bodies. But if you didn't know this, the uncanny happenings would keep you turning the pages in suspense.

The novel was written for magazine publication and appeared in three parts in 1954 under the title *The Body Snatchers*. It retains that rather specialist style of writing aimed predominantly at young 1950s' males where women were apt to cry on manly shoulders like helpless moons, yet were also heroic when called upon to be so. It's a male adolescent's ideal of the female and the author, who was in his early forties when this was written, was presumably following convention. There are parts in the plot that don't ring true; not its fantastic parts but its everyday situations and the motivations of the people involved. For instance, after the hero has found inert humanoid forms lying in the basements of two separate houses, he allows himself to be convinced by a psychiatrist friend who has been taken over and replicated already, that his mind was simply playing tricks and the alien forms weren't there at all. Who would fall for that one? The whole thing is rather lunar.

However this wasn't supposed to be a deep literary masterpiece, it was a creepy suspense story and in that it succeeds, racking up the tension further with the hero's growing uncertainty as to who is real and who is an imitation. The ending of *Body Snatchers* differs from the left-in-the-air filmed versions. In the book the alien seed pods literally do leave in the air, suddenly quitting the Earth one night and rising back to the skies having realised it was too much of a struggle trying to monopolise the entire planet with fighters like the hero in just one small town willing to resist to the death. Those humans already 'taken over' continue on like harmless zombies and die naturally after a number of years. It's a neat explanation as to why only one small town was involved and how no one outside need ever know what happened – the true local survivors assuming nobody would ever believe them. In this the story plays more to the growing reports of UFOs and strange phenomena that were widely reported in the early 1950s than it does to Communist fears. In fact the beginning and end of the book makes reference to the odd Fortean happenings that are sometimes reported, like rains of frogs, fish or pebbles. These inexplicable happenings with their 'local' aspect may have been the author's original inspiration.

[Jack Finney. 2 October 1911, Milwaukee, Wisconsin, USA, time not known. Sun in Libra, Moon in Aquarius (if born after 5am).]

The moon as nature and nourishment:
The Darling Buds of May by H.E. Bates. (Originally published in 1958)
This makes you feel hungry almost from page one and the food and drink never let up. Your mouth not only waters when for instance Ma Larkin slices up a juicy pineapple, dished out with layers of thick cream, but also when her sensuous bulk or Marietta's slender beauty moves amongst the down-to-earth glories of abundant natural life under the dome of endless 'perfick wevver'.

The whole thing is like the Moon in Taurus exalted – the month of May, the joy of nature, the love of good farm food, the sound of birdsong, the earthy sensations of scents, tastes and sex, the material abundance… I didn't know at first if the author H.E. Bates was a Taurus, but I soon discovered that of course he was. (What else could he have possibly been?) And this heaven on earth that he created with the Larkins family casts a spell hard to resist. The life and values it exudes are eternal and in that sense the novel hasn't dated, but it does remain essentially a social snapshot of the time it was written, the other side of the 1950s' Angry Young Men 'kitchen-sink' coin.

The other magic of the book is the boundless optimism of its central characters. The abundance they enjoy is a believable promise that the restrictions of post-war Britain and the old class divide were finally easing. Ma Larkin is a Mother Earth figure, a full moon as fertile and abundant as the land. The eldest of her six children is 17 (and already pregnant). Pop Larkin, surprisingly for the amount he eats, is described as thin and sharp, but generosity is his middle name. The amount of fat and cholesterol, not

to mention alcohol, that this family packs away in one evening would give a modern doctor a heart attack just to contemplate, but the general joy, contentment, and infectious pleasure at what material life can offer makes you feel better just from reading it.
[H.E. Bates. 16 May 1905, Rushden, Northamptonshire, England, time not known. Sun Taurus, Moon Libra (both Venus ruled).]

The moon as the slide into darkness
Valley of the Dolls **by Jacqueline Susann. (Originally published in 1966)**
'Dolls' are drugs, tranquillizers that allow one to relax and sleep, and 'dolls' are women, particularly those in the entertainment and modelling worlds where physical beauty is an undeniable asset. This is a tale of three women from very different backgrounds whose lives coincide in New York and whose careers and private lives we follow through the late 1940s to the early 1960s. As a story about women – three women – it has obvious moon associations. We observe and feel for this feminine trio through grateful starry-eyed naïveté to the heights of worldly success to the hollow emptiness on the other side of fame and fortune – another way to see the three moon phases. Written and published at a time when the liberation of women was gaining further momentum this huge bestseller identifies and pursues many themes and dilemmas pertinent to the female in modern society.
[Jacqueline Susann. 20 August 1916, Philadelphia. USA, time not known. Sun in Leo, Moon in Sagittarius conjunct Saturn and Pluto.]

The moon as loss of reason and ultimate change
King Lear **by William Shakespeare. (Originally published c.1608)**

'These late eclipses of the sun and moon portend no good to us…'

Nor to anyone else in this famous tragedy that touches the themes of the loss of the sun (a king losing his power) and the moon out of control (a descent into madness). *King Lear* is not a novel admittedly, it's a play, but the basis of it can be seen to have lunar themes. Apart from the study in madness there is a triple female family element. The ageing king's insanity is largely brought on, or self-inflicted, through his attitude to and the behaviour of his three daughters, Goneril, Regan and Cordelia. (Three moons to his sun).

If his three daughters represent his Moon – his instincts, his feelings – we see how he has misjudged and misinterpreted them. His Moon has become two-thirds destructive. Only the youngest daughter, Cordelia, is his saving grace, but he banishes her.

'This cold night will turn us all to fools and madmen.'

For Lear, madness and understanding grow side by side through the long stormy night until a complete change in his worldview comes about.

'O! How this mother swells up toward my heart; *hysterica passio!*'

In his derangement Lear appears to acknowledge his lunar side in this outburst, arguably previously denied. 'Hyster' means womb and it was once believed that women became hysterical because their womb was wandering through their body.

Some other quotes from the play:

'The little dogs and all – Tray, Blanch and Sweetheart – see, they bark at me!'

Lear is imagining he is calling his three lapdogs, which from their names must be female – another link in his fevered mind perhaps with his three daughters (three bitches), only one being his sweetheart.

There are several astrological lines in *King Lear*. For example:

'by the sacred radiance of the sun, the mysteries of Hecate and the night, by all the operation of the orbs from whom we do exist and cease to be, here I disclaim all my paternal care...'

And particularly the much quoted:

'This is the excellent foppery of the world, that, when we are sick in fortune (often the surfeits of our own behaviour) we make guilty of our disasters the sun, the moon, and stars.'

One night in September 1967 at a time when Cancer was rising (29 September 1967, 10.30pm, London) the Beatles recorded the nonsense song *I am the Walrus* with a random radio performance of *King Lear* taking place in the background. Although not intended to be connected, the whole song is a nonsense dirge reminiscent of the character of the Fool in Shakespeare's play. One of the continuing lines of the chorus speaks of 'Egg Men'.

[William Shakespeare. Traditional birthdate: 23 April 1564, Stratford-upon-Avon, England. Sun Taurus, Moon Libra]

THE BLACK MOON

La Lune Noir

'One maid in petals, one in wreaths
On summer hillsides did he see
Dark visions of his Lady White
In vales stretched sunward to the sea'
[Raul Athelny]

The Black Moon Lilith

For men she is the deadly fascinating inner female, savagely shadowy and unconquerable unless accepted as a bride to enlightenment. For women she is the fiercely independent option that will not bend to any dominant authority, male or female. She is a link to the pain and beauty of nature. She knows the secret name of every God and Goddess.

We have already established that nine is a number that can be associated with the moon (three times three) and it is a number especially relevant to the Black Moon as this has a nine-year cycle.

Let's remember that the black moon *is* the moon, or is derived from it. It is the measurement of the moon (the white moon) at its apogee – the moon's farthest monthly distance from us in its slightly elliptical orbit around the Earth. Like the moon's nodes the black moon is a mathematical point, not something you can actually see in the sky.

All the qualities of the black moon must therefore be refinements of the moon itself, or based on them. As the black moon is the moon at its slowest speed and farthest distance from the Earth it represents the darkest unknowable parts of the moon, where there is little natural shelter or dependable security. This is pure blackness, where the moon is free to connect with the enormous bulk of the night sky. The black moon doesn't lighten and change like the white moon. It is not reflective like the white moon, so it is much more independent. It is always dark – and invisible. It is not a touchable body. It is a lunar ghost.

It's still a moon though and it is still feminine. An independent feminine. A dark chaotic feminine force outside the usual boundaries. Little wonder it has been linked to that dangerous, uncompromising, sexual, liberating archetype 'Lilith'.

The Moon's orbit

The elliptical orbit of the moon around the Earth is exaggerated in the diagram below. An ellipse has two focal points or foci and in this case one focus point is the Earth while the other is an invisible point in space. When we view this point in a straight line from the Earth it would cross the moon's

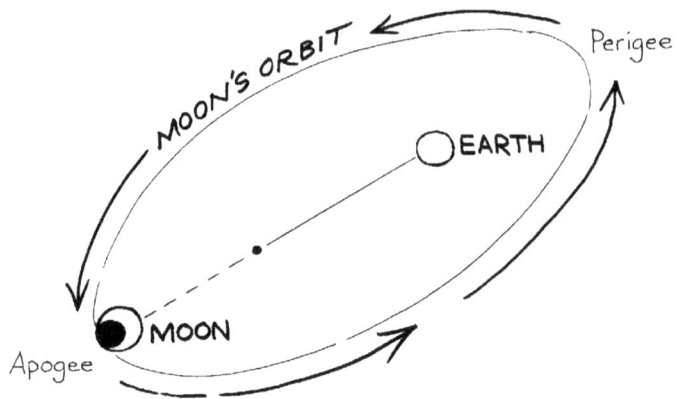

orbit at its furthest distance from the Earth, its apogee. The moon itself reaches this apogee once a month (a lunar month) in its orbit, but because everything in space is moving too, the ellipse itself gradually shifts around. It means that the moon's apogee moves about 40 degrees of longitude a year against the sky – from our view on Earth. This is the zodiacal measurement of the black moon, which takes almost nine years (8 years 10 months) to make a complete revolution through the twelve zodiac signs. If you were born when the moon was at its monthly apogee you would have both the moon and the black moon in the same place on your chart.

The black moon in relation to the white moon
If the black moon on a chart is close to the white moon is signifies that the moon is near its apogee and no matter in what sign this occurs nor in what moon phase this may be, for it will differ each month, it must be doubly blackmoonish in its meaning. Representing the white moon at its most distant and least tied to earth and its tides (the highest tides are at perigee); its emotions are independent and harder to reach. It is cut off from 'home' but open to another reality. It's Lilith rather than Eve. This is the basis for the interpretation of the black moon in any position.

Once a month when the moon is at its apogee it behaves like the black moon. If this coincides with a full moon we might expect that lunation to be more wilful and independent than usual.

If the black moon on a chart is opposite to the white moon then the moon is near its perigee, its closest point to Earth. As such it should be more earthy and realistic, slightly more conventional and down-to-earth. In other words the moon will be conjunct the point named Priapus, the earthy garden gnome.

If the black moon is square to the white moon it stands between Lilith and Priapus, trying to connect them and make something of them. If it is trine or sextile, this connection is more natural.

A Black Moon Eclipse

If an eclipse of the sun takes place at the point of the moon's apogee it is called an annular eclipse. It is not a total solar eclipse because the moon's disc being at its farthest distance from the Earth is slightly smaller than the sun's disc and cannot completely cover it, even when aligned precisely. The word 'annular' means 'ring' and describes the ring of light that remains around the moon's silhouette at such an eclipse.

An eclipse at the moon's perigee, the opposite point, can certainly be total because this is the closest distance of the moon to Earth and so the moon appears slightly bigger and can block out the sun. The moon (the white moon) is much more in command with a solar eclipse at its perigee.

At apogee the white moon can never quite be in command in the same way. Even at an eclipse it cannot completely obliterate the sun and this reveals again the subtle difference between the black moon and the white moon. The black moon is not totally lunar – in the sensitive, secure, maternal sense. It's tied to the moon but it's more independent.

A similarity to the Moon's Nodes?

The black moon and the moon's nodes are both moving abstract points in space. We can't see them through a telescope. The key difference between the black moon and the moon's nodes, in an astronomical sense, is that the nodes are points derived from *Sun, Moon and Earth* while the black moon is derived only from *Moon and Earth*. The Sun is out of the equation here.

An important feature of this is that unlike the moon's nodes, the black moon can – and does – go out of bounds in its measurement. It travels beyond the ecliptic in its declination and does it regularly. It can also go out of bounds when the white moon cannot, although it is in essence tied to the moon's major and minor standstill cycles.

When the white moon is out of bounds, the black moon will usually be also. But where the white moon can travel up and down in declination, going out of bounds in both the northern and southern hemispheres in the course of one lunar month, the black moon will remain either in the north or the south all the time. The black moon in declination follows its nine-year pattern with roughly half of that time (four and a bit years) in one hemisphere and the rest in the other. Naturally it will not be out of bounds all of this time but will move gradually up and down, through higher or lower declination degrees, crossing the equator twice in nine years.

An interesting point about this is that only some white moon out-of-bounds periods coincide with the black moon out-of-bounds periods, and therefore when both are out of bounds at the same time the energy will surely be wilder than others.

As with its out-of-bounds periods the black moon hovers for quite an unbroken while around the zero-degree area of the equator, fluttering back and forth across this line for about six months before it finally heads off in its underlying direction, either north or south. Because of the Aries Point nature of the zero-degree line the black moon can possibly manifest in a more earthly way when it is here, presenting itself more clearly and making its meaning understood.

The meeting point of north node and black moon by longitude occurs every six years. If we turn back the clock and view some recent conjunctions we find the following:

In September 1997 the declinations of north node and black moon were conjunct and also centred around solar and lunar eclipses. This period can immediately be identified historically – and in a lunar way – as coinciding with the death of Princess Diana.

The next meeting point of node and black moon by longitude was early September 2003. The north node and black moon were conjunct by declination but there were no eclipses.

The following meeting point by longitude was September 2009. Node and black moon were conjunct by declination, but again there were no eclipses.

In September 2015 the north node and black moon meet by longitude and declination – and there are eclipses.

There appears to be no general pattern (unless it's a vast one) as to when eclipses coincide with the north node/ black moon meeting point, although the declinations of north node and black moon do regularly coincide with their longitude meeting points. But they also meet in declination in-between times, midway between the regular six-year dates, and this can sometimes (but not always) be the meeting of black moon with the *south* node.

Priapus

Priapus is the name of the point directly opposite the black moon in zodiacal longitude. It was named after the ancient Greek god of orchards and gardens, the phallic Priapus.

If Lilith in mythology has some link with Eve as the first woman, Priapus the god of fertile gardens can be seen as having a loose connection with Adam the first man. Adam was essentially a gardener. 'God put the first man in the garden of Eden to dress it and keep it' (Genesis 2.15). Priapus, whose fertility is emphasised by his huge erect phallus, is rarely found in the statuary department of your local Garden Centre, although the popular garden gnome can easily be seen as an acceptable modern equivalent. (Scarecrows may also be related to him). In earlier times in Mediterranean climes statues of Priapus in all his ugly glory were commonly found in vineyards and gardens to promote the fecundity of the earth.

In Norse mythology Freya, the Queen of Heaven, was often pictured as riding a cat or in a chariot drawn by cats. She wore a precious necklace, *Brisingamen*, made by dwarfs. To the revulsion of Odin, Freya agreed to sleep with four dwarfs in turn as the price for this fabulous necklet. There is an echo here of the sexual connotations in the story of Lilith where she consorts with outlawed spirits to the annoyance of God and Adam. Coincidental too is that Freya's twin brother Freyr, a masculine form of Freya's force of fertility and beauty, is often identified as the Norse equivalent of the Roman *Priapus*.

Priapus is the point of the moon's perigee, its closest distance to Earth, therefore more closely tied to the earth than Lilith.

Lilith: the history and legends
The history

Several narratives vie for a place in the origin of astrological Lilith…

One story goes that back in the seventeenth century the Italian astronomer Giovanni Battista Riccioli (1598-1671) ordered a new lens for his telescope. Riccioli, who was professor of astronomy at Bologna, was pleased with the extra clarity he obtained and was surprised to find a small body, previously unseen, that hovered near the moon. Because of its lunar proximity this intriguing new satellite was difficult to track and didn't always appear, but Riccioli persevered and eventually worked out an ephemeris for it. The fact that this dot of light was sometimes visible and sometimes invisible may have added to its mystery but Riccioli had a good reputation;

he had mapped and given the appropriate names that we still use today for the 'seas' on the moon's surface, and his calculations were accepted by his peers.

We presume it was not long after this that the awful realisation dawned on Riccioli that perhaps his marvellous scientific breakthrough was not what he originally thought. A slight technical flaw in his new lens meant that a point of refracted light appeared when the telescope was pointing at a certain angle. So in fact he hadn't discovered anything, except that he had a dodgy telescope.

Riccioli duly informed his colleagues and the whole thing would have been forgotten had it not been for a surviving copy of his ephemeris calculations that turned up three hundred years later in an attic in France. Robert Ambelain, a keen scholar of the Kabbala, seized on the document and added some esoteric interpretations of his own. As is natural when unearthing a rare old manuscript, a tendency to believe it to be of lost ancient wisdom is not surprising. So when Ambelain republished it as a book in 1927 called *Lilith*, replicating and continuing Riccioli's ephemeris, he was unaware that it was originally no more than a figment of faulty equipment. However it was he who saw the lunar connection with the occult figure of Lilith – he who named it 'Lilith'– and the esoteric undertones were now strongly represented.[1]

Ambelain's work found interest in the USA and Ivy Goldstein Jacobson later published her *Dark Moon Lilith in Astrology* possibly using Ambelain's ephemeris and certainly her own astrological expertise. By now it was more or less accepted that the astronomical body called Lilith was either too small to be clearly identified or was supposed to be an invisible force. Kelley Hunter suggests that it may be a moon 'that used to be'.[2] Ambelain's revelation and ephemerides had a longer run in America than France because French astrologers were experimenting with that other lunar point, the outer focus of the moon's orbit called *la Lune noire*. This calculation was credited to the work of Dom Neroman (Maurice Rougié) in 1937. Over the following decades of the twentieth century the esoteric meanings of the

---

1. Information on Riccioli's false discovery and Ambelain's rediscovery in the 1920s from Marc Beriault's article 'The Dark Moon: The Mystery of Lilith', *Considerations*, New York, May 1997.
2. M. Kelley Hunter, *Living Lilith: Four Dimensions of the Cosmic Feminine*, The Wessex Astrologer, Bournemouth, 2009. For an extensive account of Lilith in all her mythological and astrological manifestations this book is highly recommended.

Dark Moon Lilith began to fuse with the Black Moon, *la Lune noire*, – now the most accepted point for this dark and mysterious force.

However the Ambelain story is only part of a somewhat confusing larger picture, as the astrologer Sepharial, who was greatly influenced by the French system of esoteric astrology, had named a lunar object 'Lilith' back in the nineteenth century. The discovery of this second moon orbiting the Earth is credited to Georg Waltemath from Hamburg with whom Walter Old (Sepharial) was in correspondence. Waltemath claimed to have discovered several extra moons but it was the main one of these that Sepharial worked with, producing an ephemeris and naming the moon Lilith in 1898.[3] Sepharial invented a glyph for Lilith that he used on charts; a circle with a line through its centre, and he was most definitely speaking of this as 'Lilith' in his surviving notebooks. According to his biographer Kim Farnell, Sepharial believed Lilith to be an actual moon of approximately the same size as Luna, so black that it was invisible except when in opposition to the sun or transiting across it. Like many in his day Sepharial was keen on biblical prophecy and 'Lilith' was an obvious name for this dark moon which in his view promoted swift changes and upsets. He also wrote, in a magazine article in 1898, that the influence of Lilith was like the Moon's, but more violent and unfortunate.[4] He believed it to move about three degrees a day.

The aforementioned Ivy Goldstein Jacobson confirms in the first pages of her book *The Dark Moon Lilith in Astrology* (1961) that Sepharial was the first to name and use the dark moon, and it now appears she may have followed Sepharial's ephemeris rather than Ambelain's.

The whole idea of a second moon was rife in the 19th century not least as it is mentioned in Jules Verne's *From Earth to the Moon* (1865) in which the space travellers see a smaller second moon passing their craft. This was based on the actual claim of Frederic Petit, director of Toulouse Observatory, who discovered a second moon in 1846.

None of these, apart from Neroman's second focal point calculations, are what is now termed the 'Black Moon'.

---

3. Kim Farnell, *The Astral Tramp*, Ascella, London, 1998
4. Sepharial writing in *Coming Events*, April 1898. Information by Kim Farnell in *The Astral Tramp*, see note 2.

The legends

From a disjointed hotchpotch of ancient legends and mistranslations comes the surviving story of Lilith – the first wife of Adam. She descends to us via Sumerian and Hebrew mythologies, though is not mentioned in the Bible story of the first man and woman. Nevertheless this is where she was believed to have made her appearance, among that fruitful garden paradise 'eastward in Eden'.

It has sometimes puzzled readers that before Adam and Eve are specifically mentioned in the King James Bible it says this:

'So God created man in his own image, in the image of God created he him; male and female created he them.' (Genesis I: 27)

'God created man in his own image' does not mean God created *a* man in his own image, it means God created mankind, men and women, in his own image, and 'male and female created he them' suggests that both sexes came into existence together. This seems in contradiction to the directly following Genesis story of Adam and Eve, where Eve is some kind of afterthought created from Adam's rib. 'Male and female created he them' may be part of an older text that slipped through the early censor's net, where in common with most other creation myths the original 'God' was both male and female. In another early line (Genesis I: 26) God says, 'let *us* make man in *our* image, *our* likeness' (my italics). Not *my* likeness. 'God' was a translation of *Elohim*, which is a plural word encompassing male and female. Not only did the first human male and female appear together but they were created by a divine mother and father – together.

The *Ozar Midrashim*, surviving from 7th-10th century AD, documents the old story of Adam and Lilith who were created as equals but from the start didn't get on well.[5] Lilith would not 'lie beneath' Adam, and Adam continued to insist that she should, implying that her place was beneath him in all ways. Lilith argued that they were equal because they were both formed from the earth, but their disharmony intensified. Eventually Lilith had had enough and she called out the secret name of God and flew into the air. Adam also then had a dialogue with God – although it seems he didn't know his 'secret name' – and told him that his woman had run away. God, who didn't appear to know this until Adam told him, immediately sent three angels to bring her home.

---

5. The following account is based on a Ethel Vogelsang's documentation of the myth in *The Relevance of the God-image in the Creation Myth of Lilith and Adam* privately published by The Guild of Pastoral Psychology, Guild Lecture 252, © Ethel Vogelsang 1995.

These angels are named elsewhere as Sanoy, Sansonay and Samanglof. They may be the Nephilim or 'Watchers' that guarded the Garden of Eden. These Watchers we are told later mated with the daughters of men producing a race of giants, and it was these giants, or some of them, who ran out of control causing chaos and were wiped out in the Great Flood.

Some of the giants had inherited the knowledge of their fathers, the Watchers, and taught humans specific skills. Amongst these giants, according to *The Book of Enoch*, Baraqijal taught astrology, Shamsiel taught the signs of the sun and Sariel the course of the moon.[6]

The original three angels found Lilith and gave her God's ultimatum. In as many words: 'either come back to Adam and behave yourself or one hundred of your children will die every day'. This threat cut no ice with Lilith and she told them to get lost.

The angels then informed her they would have to drown her in the sea.

Lilith told them that it was her purpose to harm all infants. Boy babies would be vulnerable for their first eight days and girl babies for their first twenty days, unless they were specially protected by the names on a kamea (magical talisman) when she would honour that promise and heal them. Otherwise a hundred of her sons would die each day. One hundred of her 'demons' would die.

This story of Lilith and Adam as it has come down to us here doesn't make a great deal of sense (to put it mildly). For instance where does all this talk of children suddenly come from? If Adam and Lilith were supposed to be the first ever human man and woman on Earth, neither of them would yet know what a child was, and it doesn't say that they had any children together. So what kind of condition was it to say to Lilith that a hundred of her children would die? What children? She has no children, and if she has who is their father? Clearly they are not human children – not Adam's children – and the translated word is shown to be closer in meaning to spirits or ghosts.

Other versions attempt to fill in a missing gap by saying that when Lilith departed from the Garden of Eden she consorted with spirits and bore demonic children. But who are these evil spirits that she consorts with? Are they the Watchers? In the creation story so far we are told that God had created heaven and earth and sun and moon, beasts of the field, fish of the sea etc. and then Adam, or Adam and Lilith. It says nothing about any other forms of spirit life floating about. Who created them then?

---

6. Philip Gardiner and Gary Osborn, *The Shining Ones*, Watkins, London, 2006

It would seem that this 7th-10th century version is a pastiche of a much more ancient story, with many pieces missing. Whatever we're left with here, its half-remembered facts about Lilith seem to have been perverted beyond reason.

Another bizarre religious notion, admittedly of interest mainly to Christian scholars and probably enforced if necessary by the Inquisition, was that the ecliptic is tilted as a result of original sin.[7] Presumably Eve and Lilith were blamed for the basic mechanics of our annual journey around the sun. The implication that changing seasons are ruled by the feminine does have some sort of logic because you can argue that it is not the masculine sun that determines the fertility of Planet Earth but the way the feminine 'mother' Earth is tilted to it. Quite how the female could be blamed rather than thanked for this is hard to get your head round.

So fertility, along with the obliquity of the ecliptic, is the province of Mother Earth. The very ancient story of a goddess learning and using the secret name of the sun god to gain power from him may have some relevance to this. In Egyptian mythology Isis learnt the secret name of Ra and healed him in the process; Lilith also knew the secret name of God, which neither Adam or Eve did. Did she then heal God, or could have done if given the chance? This is a thought-provoking proposition.

The Dark Goddess of Nature

Duende

In the early 1930s the Spanish poet Federico Garcia Lorca delivered a lecture in Buenos Aires on the 'duende'. It remains the most famous dissertation on this long-held but difficult-to-put-into-words concept – the mysterious force that springs spontaneously out of instinct. But *duende* is not only for those who carry the blood and death of the bullring in their cultural DNA, it is as universal and as easy (or difficult) to understand as the black moon.

Artistry that has the dark vital spirit of blood and mystery reminds us of the qualities of the black moon. Artistry that is flat, safe and sweet doesn't have it. Artistry with emotion may not necessarily have it.

It is a tempestuous untapped power, one of the most ancient known to humankind, but often shunned, avoided or mistrusted.

---

7. Quoted by Keith Hutchinson in *Medieval Heliocentric Universes: a Novel Perspective on an Old Problem*, Conference on the Inspiration of Astronomical Phenomena, Magdalen College, Oxford, 2003.

Lorca was at pains to point out that he was not describing the muse or the guardian angel but another thing again. As always we may be missing something in translation but it seems to me that the *duende* could in many cases be a dimension of the personal muse or angel, just as the black moon is a dimension of the Moon.

Cats

> 'When have I last looked on
> The round green eyes and the long wavering bodies
> Of the dark leopards of the moon?'
>> [W.B. Yeats. 1865-1939. From 'Lines Written in Dejection' (1915), collected in *The Wild Swans at Coole*]

Lilith was a woman who turned into a leopard at night, in George MacDonald's Victorian fairy story *Lilith*. And in historical reality the 'Adam and Eve' Lilith was often pictured with the body of a cat – as in the Queen Mary's Psalter, an illustrated 14th century manuscript.

Looking up more of these old medieval bestiaries I found that the leopard was once believed to be the result of an adulterous mating between a lion and a pard (a black panther). In heraldry a leopard on a coat-of-arms indicated that the original arms bearer was born out of wedlock or from an adulterous union. The leopard's black spots were the evidence of sin.

It explains a connection in the medieval mind between the leopard and cats in general, and Lilith who was blamed for leaving Adam in the Garden of Eden and cavorting with demons. But leopards have a much broader symbolic history. Their skins for instance were worn by Egyptian priests to show they had gained power over Set, the god of darkness and evil spirits; and there is a Jewish legend that says that Adam and Eve used leopard skins to cover their nakedness.

The familiar Ancient Egyptian symbol of an Eye with its extended lines and teardrop pattern followed the shaded markings around the eyes of falcons and big cats like the cheetah and leopard. The right eye of the falcon god Horus was associated with Ra, the sun, while his left eye was associated with Thoth, or the moon. Set stole the moon-eye and split it into parts, which were later restored by Horus and other gods. Restoring the damaged eye became a lunar ritual tied to the moon's cycle. As a sacred symbol of regeneration and as a protective amulet the Eye of Horus refers to the moon-eye.

The cat has a good case for black moon rulership. In western astrology cats were traditionally said to be ruled by the Moon or Saturn because they were nocturnal animals who could see in the dark, like an owl – also associated with Lilith,[8] but no astrological authority ever seemed completely sure about it. Moon-wise the cat is associated with the female; the word 'feline' has a similar ring, and witches were popularly pictured with cats. Like the moon the cat is uncanny; it has a sixth sense, its eyes change shape from slits to full black rounds and back again and it sees things from other worlds and times. (I once knew a cat whose favourite place to sit was on the back of a sofa facing a wall. Some years later its owners discovered that the wall used to have a window that overlooked a garden.)

Partially because of this magical quality and the association with heathen goddesses, like the sacred cats of Ancient Egypt, cats in general were feared by many later patriarchal religions. It is on record that in 1484 Pope Innocent VIII decreed that all cats should be killed.

Cats were never persecuted by the Islamic religion because the prophet Mohammed is said to have personally held them in high regard. The crescent moon symbol of Islam again seems apt.

Everyone has odd stories about the unearthly nature and fearful symmetry of cats, and fiction is full of them. Lewis Carroll's *Alice Through the Looking-glass* begins with Alice holding a black cat up to a mirror before stepping through into the Looking-glass House. ('Let's pretend there's a way of getting through into it, somehow, Kitty'.) There's a famous cat in Alice's first adventure too. The Cheshire Cat in *Alice's Adventures in Wonderland* who appears both whole and in parts – like the moon – informs Alice, 'we're all mad here. I'm mad. You're mad'.[9]

Yet there are other things about the cat that cannot be seen as totally lunar in the conventional astrology sense. The maternal instinct of the White Moon is not overly marked in cats for instance. Cats are no different to any other animals here. They are not held to be extra maternal or family oriented; quite the reverse: they are famed for being one of nature's most independent creatures. Although they can be loving pets, they are also killers who play with their prey. Cats do not like to be immersed in water, ruled by the moon, and they have nine lives (the black moon's cycle). In

---

8. In the Celtic version of Adam and Eve, Blodeuwedd, the first woman, turns into an owl.
9. *Alice's Adventures in Wonderland* by Lewis Carroll first published 1865. *Alice Through the Looking Glass* by Lewis Carroll first published 1872.

short cats are connected to many themes that are somehow the reverse of the White Moon.

It is a cat's independence that may be one of the biggest clues. The cat's curiosity and independence overrules its need for security and this fits the black moon's qualities of being outside an existing order. Cats will not conform to others' rules unless it suits them. Cats operate under their own laws.

Sex is said to be painful for cats, but they are fertile and produce many kittens. The Black Moon Lilith has been negatively associated with sex and branded a baby-killer. Cats do not kill their babies but they do kill small defenceless creatures like birds and mice in a cruel way. This might be seen as killing babies.

'A cat may stare at a king' underlines the general belief that cats answer to no one, just as Lilith defied the Old Testament God.

Despite their avoidance of water cats are clean animals, cleaning themselves by themselves without resource to ponds or streams. (Self-sufficiency from within). They are among nature's great survivors.

Black Moon: Nakshatras

The nakshatras or moon mansions are the 27 lunar divisions of the zodiac, previously mentioned in The Computational Moon section of this book.

As these divisions are derived from the moon's movement it is the natal nakshatra of the Moon that is more important in interpretation than the nakshatras of other planets, although the position of Rahu and Ketu, the moon's nodes, are often examined too.

The lunar theme would seem to suit an interpretation of Black Moon Lilith and could be more relevant than simply reading the zodiac sign (the Sun sign) in which the black moon is based. Differences between the tropical and sidereal zodiacs are no barrier here because the nakshatras, originally based on stars, are referring to the same divisions of the sky in both systems. The *sign* involved need not be relevant; the lunar mansions have meanings of their own.

In a nutshell: The natal nakshatra of the black moon may reveal the hidden spiritual path of that person, fully activated only after one's lunar reactions have been fully understood.

Non-reaction

If the White Moon is *Reflection*, *Response* and *Reaction*, then the Black Moon is the negative of those features; in other words: *Non-reaction*.

This can take two forms. In a negative sense *non-reaction* is heartless and unfeeling, like the myths of the demonised Lilith. But in its positive sense *non-reaction* is the ability to remain still and calm when negatively challenged. It does not participate or perpetuate a negative situation; it does not bind itself to the earth to create more karma. It can even go one step further. Its non-reactive stance may allow new energies to be released of their own volition stimulating positive outcomes.

Film Noir and Fatal Women

As already mentioned in the Book Reviews section, the last line of James M. Cain's 1936 novel *Double Indemnity* contained just two words:

'The Moon'.

It referred both to the silver mistress in the sky and the dangerous female protagonist – the *femme noir* – who shares this doom-laden story. Like so many of the *film noirs* that followed on the cinema screen in the 1940s and early 1950s, the plot is narrated in flashback by a man whose involvement with a mysterious magnetic woman usually spells his downfall. Yet in the process secrets are uncovered and necessary changes made. Because of the woman, who may be as vulnerable as she is deadly, the main hero appears to be drawn to complete or participate in a series of actions whose purpose is totally at the mercy of fate.

Film noir, like the black moon in astrology, defies exact classification. *Noir* in French signifies more than dark or black. The mood of loneliness or depression is hinted strongly in expressions like *broyer du noir*, to think melancholy or dismal thoughts.

The *noir* subject is often a detective or crime story, but not necessarily, the style of the films owing much to the cross-fertilisation of European directors working in Hollywood in the mid twentieth century. The main female character might be an evil femme fatale; but not necessarily. It's likely to be set in half-lit bars and clubs of a big city; but again not necessarily. It will have atmospheric tones of light and shadow – yes, almost certainly. And it is the darkness, the 'noir', that is the key. A darkness breached by a solitary desk-lamp in a midnight office, a distant neon slanting through curtain slats, or moonlight itself. *Noir* has all the hallmarks of that shadowy presence, the Dark or Black Moon.

The *noir* films stood apart and at odds to the colourful escapism of the majority of Hollywood's great audience-pleasing movies of the 1940s and 50s. Yet in a peculiar black and white way *film noir* was not depressing. It had complexity, depth and glamour.

Complexity, depth and glamour
This is a good description of the qualities of the black moon itself, especially the root of the word 'glamour' which means a kind of otherworldly enchantment.

The first use of the term 'film noir' to specifically describe the spate of American black and white crime films released in France at the end of the 2nd World War and Nazi censorship, is credited to the Italian born French film critic Nino Frank (27 June 1904, Barletta, Italy) in the film magazine *L'écon francais* (The French Screen) in August 1946.

The five films he reviewed and described thus were:

John Huston's *The Maltese Falcon* (1941)
Billy Wilder's *Double Indemnity* (1944)
Otto Preminger's *Laura* (1944)
Fritz Lang's *The Woman in the Window* (1944)
Edward Dmytryk's *Murder, My Sweet* (1944)

Sunset Boulevard **– Black Moon parallel Venus**
Another well-known *film noir* is Billy Wilder's *Sunset Boulevard* (1950) starring Gloria Swanson as Norma Desmond, a tragic aging Hollywood movie queen. It was a controversial film because it exposed in fiction a story no stranger than fact about the darker side of Hollywood's Dream Factory and star system.

Sunset Boulevard was first released in the USA on 10 August 1950 and in the absence of an accurate time our chart (overpage) is set symbolically for sunset at the place the film describes: Sunset Boulevard, Hollywood, Los Angeles, California. (34N06 118W20. 19.41.52 PDT)

The White Moon in Cancer is out-of-bounds and waning – in the story the house on Sunset Boulevard exists in an age gone by, independent of convention, the grounds are overgrown and run to seed, a powerful woman denies the outside world and lives on past glories.

The Sun in Leo is conjunct Pluto – the film is narrated by a dead man, a young writer who fatefully stumbles into this opulent Gothic mansion. The house's owner, the screen goddess of yesteryear, is obsessed with her self-image and will descend into madness as the Leo Sun is pulled down into Pluto's underworld. The Sun conjunct Pluto is scorpionic and when Norma Desmond states at one point that she is a Scorpio, we can well believe it. She dresses predominantly in black; she hides her hypnotic eyes behind dark glasses. She is passionate and intense, living and loving in extremes,

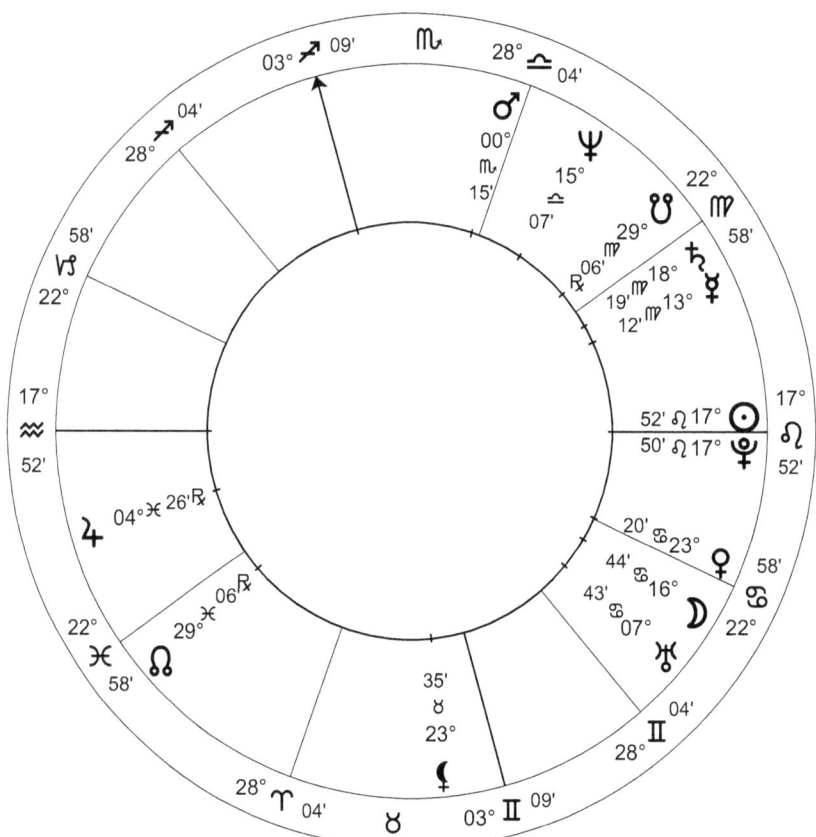

Declinations:

Moon: 27°N16' (out of bounds)
Uranus: 23°N30'
Pluto: 23°N09'
Black Moon: 21°N51'
Venus: 21°N28'
Sun: 15°N29'
Saturn: 06°N24'
Mercury: 06°N17'
South Node: 00°N22'

North Node: 00°S22'
Neptune: 04°S30'
Jupiter: 11°S04'
Mars: 12°S13'

attempting suicide when her feelings are hurt and ultimately killing her lover when he crosses and spurns her.

By declination the black moon is parallel Venus, and sextile to it by longitude – the character of Norma Desmond has understood her independent femininity only through her beauty and glamour. Along with charm and good looks Venus awarded her wealth, freedom and privilege.

The black moon is in nakshatra number 3. Krittika, the Knife, a lunar mansion ruled by the Sun. The positive hidden path of Krittika is to be heroic and sharp, to act with authority and determination, which in this story the woman does. Inspired by the young man who has entered her life and fired by the belief that she can make her movie comeback, Norma takes command of her situation. Defying her age she beautifies herself in readiness for the film studio's call. But Krittika has a malefic side that can be destructive when other sign-posts are ignored.

Before the Black Moon can be fully handled (on any chart) the White Moon must be understood. The White Moon in the chart of *Sunset Boulevard* is strong in its own sign of Cancer but is closely square to Neptune. The biggest flaw in the character of our dark heroine is the tendency to live in a fantasy world of the past. Unhelpfully the deception that she is still a big star is deliberately maintained by those closest to her.

When Norma Desmond descends the stairs into madness in the final scene, believing the flashing bulbs of the journalists and the excited gathered crowd are part of a film in which she is starring, we meet again the light-versus-dark dilemma of this chart. Norma was a Scorpio inhabiting the dark who lived for the call of 'Lights! Camera! Action!' Noticeably 'lights', for only these could bring the make-believe person she had become, to life. The film fades out with a close-up on her face, symbolically blending black and white into a devitalised grey, gelding the vital Scorpio extremes of dark and light into a non-entity, blunting the black moon in Krittika to an ineffectual force. For this once beautiful actress the secret name of God was 'Venus' (Black Moon parallel Venus) but she did not understand her own Moon and its subtle Neptunian deceptions. Her lover had gone out with a shot from her gun, but Norma's face would gradually fade, just like the dark Moon waning to oblivion.

CONCLUSION

> ... all ye that walk in Willow-wood,
> that walk with hollow faces burning white...
> [Dante Gabriel Rossetti, *Willow-wood*, from *The House of Life* 1881]

We have now almost come to the end of our walk through Willow wood with all its lunar faces and phases. We have seen that the moon can be a beautiful lady, a crone, a wise priestess, a siren, a witch and much else besides. We have seen she is both the Virgin Mary and the demon Lilith. We have seen it and felt it. She hides then shines then hides again, she grows and fades and then she dies. It only remains to discuss the correspondences and then we are done.

Why are they ruled by the Moon?

> 'All things by immortal power Near or Far
> Hiddenly to each other linkèd are
> That thou canst not stir a flower
> Without troubling of a star'
> [Francis Thompson 1859-1907]

In the following correspondences both ancient and modern of flora and fauna and objects and places ruled by the Moon, our investigations are not so much concerned with 'who said so' (which authority decreed or first noted the lunar rulership – if such a record were possible) as 'why in popular thought might it be so'.

Traditional lists of items that are ruled by the Moon can run to many pages and include flowers, professions, cities, illnesses etc., some of which have become obscure or less important as the centuries have rolled by. Updated variations on the same old themes continue to take their place and everything on earth finds its rulership in the sky. But to my knowledge very few sources attempt to tell you why such-and-such a thing is supposed to be ruled by a particular planet or, in our case, the Moon. Much is self-explanatory; simple chains of association like Milk – Breast – Motherhood

– Cancer – Luna, are more than obvious. But it's worth being reminded of some of the many components that make up a common lunar thread, for who but an astrologer might see the hidden connection between a crab and a cucumber?

Energy resides in everything both animate and inanimate, so that when we say the energy of the Moon resides in silver (or that silver is ruled by the Moon) we mean that some sort of vibration is present in silver that operates on the same level as it does in a pumpkin or a piece of chalk. In practical terms it means that many things are symbolically interchangeable, being on the same wavelength as each other. "Thou canst not stir a flower without troubling of a star". If we want to attract the influences of the Moon, we can visit places or surround ourselves with items symbolic of it.

Following is an alphabetical lunar list…

Baskets are ruled by the moon because they usually contain food. They are concave or nest-shaped and are made of yielding substances, woven or plaited. A basket of eggs is a classic moon symbol.

Boats or ships are ruled by the moon because they are protective enclosed little worlds floating on the element of water. They are traditionally given female names. If a vessel is particularly crescent-shaped, like a canoe, or bears life like a cradle or ark, its relationship to the moon is even more obvious.

Soft white crumbly stones are moon-ruled so **Chalk**, which is formed from sea shells, is an obvious contender. Chalk has many associations with childhoods of the past. Lessons were chalked on a classroom blackboard; jumping and skipping games were chalked in the playground. In pregnancy some women crave chalk. Chalk is porous and contains calcium, like milk, the moon's liquid. Chalk and cheese are not so different after all.

And **Cheese** is ruled by the moon not only because it is made from milk but because the moon in the sky can look like a round cheese, and in popular nursery rhyme thought was actually made of green cheese.

Chocolate is ruled by the moon because it is essentially a dark liquid drink. Later it was manufactured in solid forms. It comes from seeds inside the pods of the fleshy cacao plant and originated in Central and South America where its name meant 'bitter water'. Today it is often regarded as a comfort food to which women in particular can be addicted.

Cobwebs are soft meshy veil-like entanglements found in old houses and abandoned places, and therefore moon ruled.

Coral is ruled by the moon because it is a jewel of the sea. It was once thought to be protective to children and in some cultures parents still give gifts of coral to their offspring for this reason. Corals breed by releasing reproductive cells at full moon. Nicholas Culpeper in his *Complete Herbal* recommended adding coral to breast-milk in grains to prevent falling-sickness in a baby and convulsions in later life. He also listed it as a remedy to 'stop the immoderate flowing of the menses'. Coral is a girl's name.

For **Crab** see **Shell fish**.

Cucumbers are ruled by the moon because they are cold moist succulent fruits of the earth, and they are vaguely crescent shaped. Culpeper the herbalist recommended their juice as an antidote to sunburn. Because of their cool succulence many other salad ingredients are moon ruled, like Cress, Lettuce, Tomatoes etc.

The name **Cynthia** derives from one of the ancient names for the moon goddess.

Dew is ruled by the moon because it was once believed to fall overnight from the moon, and has been known as Selene's child or the moon's daughter. Washing in May dew (the month in which Luna is exalted) brought feminine beauty and protection from witchcraft.

The Greek goddess **Diana** (Artemis) is one of the main surviving identifications of the moon goddess in western culture. She is commonly depicted carrying a bow, which is crescent shaped. The New Testament of the Bible mentions that the worship of Diana was strong in the city of Ephesus, appropriately a centre for silversmithing. [Acts of the Apostles 19].

Ducks are ruled by the moon because they float on water like boats. They have always been children's favourites, floating in baths, quacking in cartoons, waddling like toddlers. To 'duck' is to avoid a missile or impediment rather than meet it full on. The soft feathers of the duck are used for pillows (eider down) to aid sleep.

Ducks lay **Eggs**, as do chickens, and country superstitions regarding the hatching of eggs have strong lunar connections. Setting an odd number of eggs under the hen was thought favourable, thirteen being the best number of all. (There are approximately thirteen moons in a year). Sunday, the day of the Sun, was regarded as the worst hatching day and so avoided. In the animal world an egg itself resembles a moon by being round, white and smooth and acting as a protective home for the forming infant. Chocolate eggs are gifts at Easter, symbolising new life and the full moon.

Another moon fowl is the **Goose** and like many animals ruled by the moon 'goose' is the feminine form of the name. In children's rhymes and stories, the Goose figures prominently. 'Mother Goose' is a common phrase. Geese were once employed as protectors of homes and wild geese are pictured flying across the moon.

A **Harbour** is ruled by the moon because it is a watery haven or home shelter for sea-going craft. Natural harbours are crescent shaped.

And the **Hare** is one of the classic moon-ruled animals whose crazy leaps of intuition and ancient fertility connections associate it with the spring and the full moon celebration of Easter. The Easter Bunny who brings eggs is a modern descendant. Hares have a reputation for being slightly loony or 'harebrained' or 'as mad as a March hare'. Artemis, the goddess of the moon, held a special affection for the pregnant hare and huntsmen in earlier times were aware of the taboo in killing one. Later, like most creatures associated with the moon and pagan beliefs, hares were regarded at the very least with suspicion. A pregnant woman (the moon) feared producing a hare-lipped child if she saw a hare by accident, sailors would not mention the name of the animal at sea, and hares were believed able to change into human form as witches, only to be destroyed with silver shot. Buddhists and Eastern countries have seen a hare rather than a 'man' in the moon. In Ancient China the hare was regarded auspiciously as the servant of the moon queen.

The human female produces eggs in accordance with a lunar rhythm and when many women live together communally it has been reported that their menstrual periods synchronise, their moons becoming receptive to one another and eventually to the moon's phases in the sky. So a **Harem** is a sacred moon place where no ordinary male could enter. It is said to have descended from the meetings of the priestesses of the moon goddess, later the Vestal Virgins, whose children were once accepted as divine rulers.

An **Heraldic Shield** is ruled by the moon because it depicts the coats-of-arms of a family tree. A shield of any kind is similar to the protective hard outer shell of Cancer the Crab, but when related to the ancestry of a family or institution it is doubly relevant. The late Sir George Trevelyan.[1] likened the trappings of a family crest in heraldry to be symbolic of a soul entering a particular ancestral stream, with the two supporters either side of

---

1. George Trevelyan in *A Vision of the Aquarian Age*, Coventure, London, 1977, and lectures on this theme.

the shield as the spiritual guides, the mantling as the etheric body and the motto inscription as an outline of the soul's task.

The flower-de-luce or fleur-de-lys was a popular heraldic symbol and Culpeper followed tradition in his *Herbal* by giving this **Iris** plant to the moon. The yellow flag iris that grows in ditches and ponds illustrates its watery connection but the iris in general is a three-petalled flower and therefore in sympathy with the Triple Goddess. In classical mythology Iris was the goddess of the rainbow providing another moon link with bows and crescents. The White **Lily** is another moon flower, associated with purity, death, and changing states.

Thick juicy **Leaves** of any kind are ruled by the moon because they contain moisture. Leaves offer a visual image of waxing and waning as they grow and change colour and fall, and the wistful beauty of autumn leaves is analogous to the Old Moon phase. There is also the poetic association of leaves renewing life on a family tree.

All soft fleshy plants are moon-ruled especially, if like the **Melon**, they are round in shape like a woman's breasts. The pumpkin is somewhat similar but has an additional association with witches and changing shape at midnight. In the story of Cinderella a pumpkin is changed into a princess's coach by a fairy godmother, a benevolent moon figure.

It was an old belief that **Mice** were manifestations of the soul, (the moon was the abode of souls), and the unexpected appearance of a mouse can still produce irrational fear in some people. Mice have a tendency to inhabit homes, creeping out at night, and have strong associations with motherhood and babies, ('the patter of tiny feet'). They are attracted to milk-based cheese. They are favourite characters in children's books too. Mickey Mouse is the symbol of a nation with the Sun in Cancer.

Milk is ruled by the moon because it is the natural nurturing liquid that passes from mother to child. All milk-based dairy products including butter and cheese are therefore ruled by the moon.

A **Mirror** is ruled by the moon because it is a symbol of reflection. Water also acts like a mirror. Household mirrors are 'silvered' at the back and can only reflect what is already there just as the moon is only visible because it reflects the sun's light, yet a different part of ourselves often seems to gaze back. Several fairy stories begin with characters entering through a looking glass into a lunar realm, while uncanny beings that shun the sun have no reflection at all. The superstition of having to endure seven years bad luck if a mirror is broken is tied to the 28-day lunar cycle.

Monday is unmistakeably named after the moon (Moon-day). In the not so distant past Monday was washday, a day of household and domestic chores to countless numbers of women. Monday was also a favourite day for ceremonies involving ships and water. As the first day of the working week after the weekend Monday is often approached with a melancholy air.
[German: Montag, Danish: mandag, Dutch: maandag, French: lundi, Italian: lunedi, Spanish: lunes]

The **Moth** is the slightly sinister butterfly of the night. Like a bat it can flutter to our face from the dark shadows and unnerve us. It is a delicate creature that flies by the light of the moon, often attracted to lamps and candles as substitutes in the darkness. One large variety of moth is actually called *Actias luna* or Luna Moth.

Mushrooms are ruled by Luna because they grow overnight and have the appearance of little white moons. Old gardening lore recommends you should try to pick them at full moon. 'Magic' mushrooms may take you into dreamland.

Like most sap trees, the **Palm Tree** is said to be moon-ruled. An oasis of palms in the harsh sun-ruled desert offers complementary gifts of shade and water, both darkness and fluids being ruled by the moon. In fact most palm trees grow by the sea. The swaying motion of palms in a wind displays their strength in yielding rather than standing rigid, where 'masculine' hardwood trees may be too brittle to stand for long against strong sea-borne winds. Coconuts, which grow on palm trees, are containers of milk.

The **Pearl** is the moon's foremost jewel because it is formed in the maternal deeps of the sea and has the cloudy iridescence of moonlight. 'Mother'-of-pearl is aptly named and one superstition holds that to wear mother-of-pearl on a moonlit night will make a woman fertile. The lustre of these tiny moons is softer and subtler than the flashing diamonds of the sun. Interestingly a pearl is a cancer (a growth). Culpeper recommended grains of pearl to 'encrease milk in nurses'.

The name **Phoebe** is another that derives directly from the name of a moon goddess.

An old black and white **Photograph** can be seen to be moonlike, especially if it is from a 'family' album, because it is a memory of the past and of ancestry. The photograph would have needed the aid of silver solutions in the dark room to come to life and its print in shades of black, white and grey, reflect the moonlit world without the Sun to give it colour.

The **Poppy** is a flower with milky juice and sleep-inducing properties and was once believed to weep tears. It is still used as a token of remembrance for the souls of the dead.

For **Pumpkin** see **Melon**.

Salt is ruled by the moon because of its colour and its association with cooking, and therefore nourishment. It was once believed that if you went without salt you would become insane, and salt was and is often used as protection in magical ceremonies. In the grip of emotion you taste the salt of tears and it is also the taste of the sea.

It was from the maternal oceans that all life on earth evolved and there is no argument but that the moon has an influence on the **Sea**; it is one of the few tangible proofs of astrology – that a heavenly body has influence on the earth. Areas on the moon were called seas or maria when they were specifically given names in the seventeenth century, almost all of which describe emotions or have astrological resonance with lunar rulership: The Sea of Serenity, the Sea of Tranquillity, the Sea of Crises (first to appear at new moon), the Sea of Fertility, the Sea of Moisture, the Sea of Nectar, the Sea of Rains, the Sea of Clouds, the Sea of Cold, the Sea of Waves…
It was always good advice to the newcomer learning astrology that to know what the moon rules, simply get acquainted with the names on a map of it.

The name **Selina** is another that derives from the moon goddess.

Because they live in the sea and have shells like Cancer the Crab, all **Shellfish** are ruled by the moon. 'Shell fish grow with her increase…' said Pliny the Elder, two thousand years ago. A crab's claws are crescent in shape lending further weight to the lunar association.

The moon is the **Silver** Goddess, the feminine equivalent to the golden Sun. These two metals, silver and gold, have been associated with the moon and the sun respectively since the earliest known times. In the moonlight everything appears silver, especially water, and like the moon silver turns black with age. People often turned silver coins at the new moon to attract luck and a gypsy fortune teller would ask you to cross her palm with silver before she tuned in to her subtle lunar vibrations.

Snails are ruled by the moon because they ooze a moist sticky trail and, like shellfish, they carry their homes with them.

Psychologically a **Spider** represents the devouring feminine aspect of the Dark Mother, the dark moon, and triggers irrational fear in many people. By mythological account the first spider was a woman, Arachne, who dared to match her skill at spinning with a goddess. The moon's invisible pull on the tides and emotions has been likened to a spider's silken strands. Not all spiders are Black Widows but the name is strangely apt.

Streams and **Rivers**, springs and fountains are all ruled by the moon because they are flowing water, or Cardinal Water like the sign of Cancer. Ditches and marshes and the rushes and weeds that grow in them are also moon-ruled.

Last but not least in our brief discourse on the moon's dominion is one of its time-honoured symbols, the **Wolf**. We met and discussed the wolf briefly when we were examining the Tarot card of the Moon. The fact that wolves bay at the moon is probably its primary identification and the belief that wolves have a strong sense of family loyalty associates them further with Luna. In legend it was a she-wolf who often acted as a foster-mother to suckle human infants, as in the story of Romulus and Remus the founders of Rome, or with Mowgli in *The Jungle Book* etc. In fairy stories the Big Bad Wolf often dresses as a woman, as he does in *Little Red Riding Hood* or he is intent on destroying homes, as in *The Three Little Pigs*. Psychologically a wolf or rabid dog is said to be a symbol of a woman's shadow side. The fabled Werewolf must be mentioned again too as this creature's appearance waxed and waned with the changing phases of the moon itself and was another of those magical animals that could only be laid to rest by a bullet made of silver.

13 Places Ruled by the Moon

Following is a very small selection of geographical places with lunar associations.

Argentina

'Argent' in Latin means silver and the river that flows through Argentina, the Rio de la Plata, means Silver River. The birth chart of its modern nation took place on a Full Moon – with the Sun in Cancer (9 July 1816). It may come then as no surprise that the tango, a highly charged emotional dance of close embrace, originated in Argentina.

Idaho (USA)

After a full moon the previous night, Idaho entered the Union as one of the United States on 3 July 1890 when the Sun was in Cancer. It has been called the 'Spud' State from its production of these lunar tubers, and appropriately the Craters of the Moon National Monument, which comprises over 80 square miles of volcanic lunar-looking landscape, is sited here. Idaho means 'light on the mountain'.

Manchester (UK)

The name of the city of Manchester, the second largest city in the United Kingdom, derives from 'breast-shaped hill'. The city boomed in the Industrial Revolution particularly with the manufacture of household linen. The Manchester Ship Canal was another great Victorian engineering feat turning Manchester into an ocean-going port. Although the town was a pioneering example of modern capitalism it also has a strong history of support for left-wing causes (Suffragettes, Marxism, Trade Union Congress etc).

Netherlands

The Netherlands, or Holland, is a country traditionally ruled by the Water sign Cancer. Statistically it is full of water, through its rivers, tributaries and canals, and because of its flat low-lying area. A quarter of the country is below sea level. The Dutch were a foremost maritime power historically with ships sailing out to trade and establish colonies across the globe. Today the nation practices an advanced policy of social welfare, ranking it high as a caring homeland where the wishes of the people (the moon) are reflected in their governmental laws. Euthanasia for example is legal – an individual's wish to die is honoured, and it was the first country to officially accept and recognise same-sex marriages. Among other things the Netherlands is one of the world's largest exporters of cheese, tomatoes and cucumbers.

New Hampshire (USA)

New Hampshire entered the Union as a State on 21 June 1788 when the Sun was in Cancer. Historically, delegates from New Hampshire were the first to vote for the Declaration of Independence on the 4th of July 1776. The New Hampshire chart mirrors the USA's in several particulars: Sun conjunct Jupiter in Cancer; Moon in Aquarius; Venus in Cancer... In fact it marks the USA's first Jupiter Return. The state flag and seal depict a ship. The largest city, Manchester, derives its name from Manchester in England, see above.

New York City (New York USA)

Cancer rises on the 1626 foundation chart of New York City, the purchase of Manhattan and culminates on the 1898 chart of the consolidation of its boroughs. Its Cancer rulership recalls the original links with Holland, with names like Peter Stuyvesant and New Amsterdam. New York is a port, surrounded by water, where a huge statue of a woman has maternally welcomed many to a new home. The Cancer Sun of the USA explains why this city remains so especially symbolic of the United States even though it is not the capital.

Further Note on New York City: When this city changed its name from New Amsterdam to New York it may have brought in a Gemini influence, because the chart for York in England, which it was then renamed after, has a Gemini Sun (18 May 1396 OS). Cases can be made for the cosmopolitan atmosphere, the many languages, the Twin Towers and for the double-sounding address 'New York, New York'. The English town of York itself has both Gemini and Cancer associations, see separate entry below.

And speaking of the United States and its Cancer Sun: This is the nation that first reached the Moon. It has a very emotional and patriotic response to events. Family, children and homeland, dynasties and roots, are eternally popular themes to its inhabitants. Jupiter's exaltation in Cancer (Jupiter is in Cancer on the USA chart) drives a nation whose natural impulse is philanthropic and who understands that survival depends on aspiration.

Scotland
Cancer is a Water sign and the water of the Scottish highlands is one of its greatest treasures, being bottled and sold now as pure water but also producing the 'Water of Life' or Whisky. (Whisky distilleries were built over natural springs). Scotland is also famous for its tartans and clans – the strength and pride of family ancestry. The Scottish Gaelic name for Scotland is Alba meaning 'white'.

Venezuela
The Sun was in Cancer when Venezuela proclaimed its independence on 5 July 1811. This was appropriate for a country whose name means Little Venice – the watery Italian city ascribed to Cancer many centuries before.

Venice (Italy)
World famous as a city built on water with canals rather than roadways. In the annual Carnival of Venice, masks and robes in the moonlight obscured the outer (solar) identity of participants.

Virginia (USA)
Named after a woman (Elizabeth I, the Virgin Queen) and sometimes nicknamed 'Mother of Presidents' and 'Mother of States', Virginia entered the union as a State on 25 June 1788 when the Sun was in Cancer. As George Washington's birthplace and with the Sun sign of its nation, Virginia reflects much of Cancer's maternal and patriotic personality. The

Pentagon, Mount Vernon and other national shrines are found in the 'Old Dominion' State.

Wyoming (USA)
Wyoming entered the Union as a State on 10 July 1890 when the Sun was in Cancer. The state flag displays an American bison (grazing bovine animals are generally thought to be ruled by the moon) and Wyoming was the home of 'Buffalo Bill'. Perhaps it is no surprise that this Cancer-ruled State also contains some prominent pointed mountains called the Grand Tetons (the Great Breasts).

York (UK)
Traditionally noted as Cancer ruled, the ancient walled town of York was built on marshy ground at the confluence of tidal rivers and is still liable to flooding. The River 'Ouse' in particular has a suitably leaky name. In comparatively modern times York had the nickname of 'the chocolate city' as two large cocoa and chocolate manufacturers, Rowntree's and Terry's had bases here. ('Yorkie' bars are aptly named).

Appendix:
How to Draw a Crescent Moon

Draw two circles on a straight line making the smaller circle 90% of the larger.
The radius CD should be 90% of radius AB.
Point C is positioned halfway between A and B.

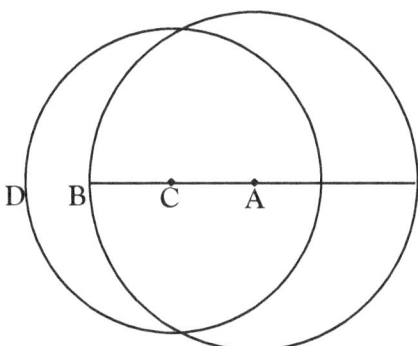

To produce a thinner crescent alter the size of the smaller circle and move its centre point (C) closer to point A.

Index

A
A Midsummer Night's Dream 127
A Vision 82
Abba 52
abode of souls 9,124
actors 27
Adam and Eve 162
Adams, Evangeline 72
Aganice 67
Age of Aquarius 111
Agrippa 88
al-Biruni 130
Alice's Adventures in Wonderland 166
Alice Through the Looking-glass 166
Ambelain, Robert 160
And Then There Were None 142
apogee 50,104,155,156,157
Arachne 179
Argentina 180
ark 88,174
Artemis 23,41,65,88,176
Assia, Lys 50
astrocartography 56,114
audience 27
Auster 11

B
Bad Moon Rising 111-114
Barrett, Syd 107
basket 174
Bates, H.E. 149
Batman 26
Baum, L. Frank 69
Beatles 151
Beaulieu 126
Beauman, Sally 134
Beauty and the Beast 30
Bedknobs and Broomsticks 72
Beerbohm, Max 117

Bergman, Ingrid 102
Bewitched 73
black madonna 25
Black Moon 50,153-171
Blake, William 118
blue moon 46
boats and ships 26,27,28,124,130,174
Bogart, Humphrey 102
Boleyn, Anne 128
Boreas 11
bow 40,41,85,175,177
bowl 26,38
Brooke, Rupert 42
Buddha's birthday 45
Buffalo Bill 183
Byzantium 39

C
Cain and his Thorns 145
Cain, James M. 143,168
canoe 174
Carroll, Lewis 166
Carter, Charles E.O. 56
Casablanca 101
cats 66,165-167
Celtic Twilight 127
Cerberus 23
chalk 174
cheese 174,181
Cheshire cat 166
chocolate 174
Christie, Agatha 142
Churchill, Winston 71
Cinderella 30, 177
Cinnamon Tree in the Moon 58
Civilisation 112
Clark, Sir Kenneth 112
cobwebs 174
coconut 178

Cold Comfort Farm 138
Conan Doyle, Arthur 139
coral 175
crab 95,175
cradle 174
Creedence Clearwater Revival 111,114
crescent moon 38-41
cress 175
Crick, Francis 49
Croft, Lara 55
crowds 48
Crucifixion Eclipse 25
cucumber 175, 181
Culpeper, Nicholas 130,175,177,178
Cynthia 175

D
dairy products 177
Dali, Salvador 119,120
dark lady 49,121
Darwin, Charles 36-38
Denslow, W.W. 69
Deucalion 27
dew 175
Diana – goddess 10,23,41,44,65,88, 95,175
Diana – Princess of Wales 15-17,34, 158
DNA 49
dogs 95,139
doppelganger 121
Double Indemnity 144,168,169
drawing down the moon 66
duende 164
du Maurier, Daphne 133
duck 175
Duncan, Helen 71

E
Easter 46,175
Easter bunny 176
eclipse – lunar 25,109-114
eclipse – solar 25,103-109

egg 7,8,120,175
egg men 151
Egypt 56,166
Elizabeth I 128,182
emptiness 57
Endymion 7,115
Erishkigal 106
Ernie 54
Eurovision Song Contest 50
euthanasia 181
Evans, Ellen 127
Eye of Horus 165
eyes 4,5,165

F
fairy godmother 177
fairy queens 127,128
fairy tales 29,64
faith, hope and charity 22
Falkner, J.Meade 138
Farnell, Kim 161
Fates 21
Fellini, Federico 33
femme noir 143,168
fertility 8
figurehead 124
film noir 168,169
final dispositor 100-103
Finney, Jack 147
first 45 rpm record 35
fleur-de-lys 177
Fowles, John 145
Frank, Nino 169
Franklin, Rosalind 49
Freya 159
Frigga 66
frogs and toads 64
From Earth to the Moon 161
Full Moon 43-56
Furies 21

G
Gala 119

Galatea of the Spheres 119
gandanta 100
Gard, Toby 55
Garden of Eden 159,162,163,165
Gibbons, Stella 138
Glasgow Airport 15
Gloriana 128
Goldilocks and the Three Bears 30
Goldstein Jacobson, Ivy 160,161
Gone With the Wind 6
Goodbye Mr Chips 140
goose 176
Gotham 25,26
Graces 21
Grahame, Kenneth 141
Graves, Robert 21
guitar 131

H
hammer and sickle 40
Hammer of Witchcraft 66
harbour 176
hare 7,43,176
harem 176
Harvest Moon 44
Hawkins, Richard 41
Heathrow airport 14,16
Hecate, 21,23,106,151
Henry VIII 128
heraldic shield 176
High Priestess 88,93-95
Hill, Benny 52,53
Hilton, James 140
Ho Chi Minh 114
Holland 181
Hollywood 143,168
Holmes, Sherlock 140
Holy Trinity 21
Hope, Bob 53
horns 38,40,96
Hunter M.Kelley 160
Hunters Moon 44

I
I Am The Walrus 151
I Ching 87
Idaho 180
Invasion of the Body Snatchers 148
iris 177
Isis 164

J
Jack and the Beanstalk 30
Jolie, Angelina 56
Jones, Prudence 66
Jung, Carl 29,43

K
Keats, John 115
Khayyám, Omar 73
King Farouk 33
King Lear 150
Krittika 85,171
Kurtz, Katherine 72

L
La Dolce Vita 33
Lady of Ephesus 88
Lammas Night 72
Laura 169
Lear, Edward 129,132
leaves 177
Lemmon, Jack 53
Leo, Alan 72
Leonids 58
leopard 165
lettuce 130
Lewis, C.S. 40
Lilith 155-171
Lilith 160,165
Lilly, William 69,127
lily 177
Little Red Riding Hood 180
lobster 93,95
Lodge, Thomas 41
Lolita 146

London underground railway 14
Lorca, Federico Garcia 164
Lot of Fortune 36,43
lunar eclipse 25,109-114
lunar month 78
Lunisequa 41

M
Macbeth 23,26,136
MacDonald, George 165
Magdalene, Mary 24,25
Magi 24
magic squares 87-91
Mailer, Norman 118,119
Manchester 181
Manhattan 181
Manson, Charles 111
Maquire, Gregory 70
Marx, Karl 40
melon 179
menstrual period 6,8,90
mermaid 122-124
Mickey Mouse 177
milk 177
Miller, Arthur 70
mirror 123,177
Monday 126,178
Monroe, Marilyn 70,119
Moon: black 50,153-171
Moon: blue 46
Moon cycle 78-82
Moon: full 43-56
Moon hours 98
Moon mansions 82-86,167
Moon: new 31-38 57-64
Moon: old
Moon: out of bounds 14,15,62,118, 157,169
Moon: phases 78-82
Moon: standstills 3,157
Moon tarot card 91-96
Moon: void-of-course 47
Moonfleet 138

moonrakers 27
Moon's nodes 105,155,157,167
moth 67,178
mother 5
Mother Shipton 68
mount of the moon 93
mouse 177
Murder, My Sweet 169
Murdin, Paul 117
muse 22,118-122
mushroom 178

N
Nabokov, Vladimir 146
nakshatras 82-86,167
NASA 10
navamsha 89
Nephilim 163
Neroman, Dom 160
Netherlands 99,181
New Amsterdam 181
New Hampshire 181
New Moon 31-38
New York City 108,181
nine 28,89,155
Nine Muses 22
1984 60
Nix 64
Noah's Ark 27
Norns 21
North Wind 11
Norton, Mary 72
Norway 52,136

O
Oberon 127
Old Moon 57-64
Old, Walter Gorn 161
old wives' tales 8,138
On the Origin of Species 35
Orionids 58
Orwell, George 59
Ottoman Empire 39

out-of-bounds 14,15,62,118,157, 158,169
owl 129,166
Ozar Midrashim 162

P
palm tree 178
Palmer, Samuel 117
Pan 10
paparazzi 33
Part of Fortune 36,43
Paschal Moons 46
Passing and Glassing 57
pearl 128,178
Pentecost 46
perigee 44,156,157
Perseids 58
Petit, Frederic 161
Phoebe 178
photograph 178
pig 131
Pink Floyd 106
Pleiades 118
Pliny 179
Plotinus 58
Pluto 10,54,106
poppy 178
Potter, Harry 66
pregnancy 8
Pre-Raphaelites 118
Priapus 157,159
Prince Charles 16
Prince William 16
Princess Diana 15-17,34,158
progressed moon 96
psychic tendencies 9
pubs 60,62
pumpkin 179
Punch 61
purdah 109

Q
Queen Elizabeth I 128,182

Queen Mary's Psalter 165

R
Ra 164,165
Raleigh, Walter 128
Rebecca 133
Riccioli, Giovanni Battista 159
Rider-Waite tarot pack 95
rivers 180
Rolls-Royce 124,125
Romulus and Remus 180
Rooney, Mickey 53
Rosalynde 41
Rossetti, Christina 57
Rossetti, Dante Gabriel 118,173
Rudhyar, Dane 78,97

S
Sabrina the Teenage Witch 73
Salem witch trials 70
salt 179
Saracen 39
Sariel 163
Saturn chasing the Moon 87
scimitar 39
Scotland 182
sea 8,130,138,160,179
Selene 8
Selina 179
Sepharial 161
Set 165
Seven Wonders of the World 88
shade 178
Shakespeare, William 121,150
shell 123,137
Shelley, Percy Bysshe 121
shellfish 95,179
Ship of Fools 27
shoreline 8,122
sigils 91
silver 125,130,179
Silver Lady 124,126
snail 179

solar eclipse 25,103-109
Solomon's temple 88
South Wind 11
Soviet Union 40,106
Spenser, Edmund 128
spider 179
sputnik 106
Star of Bethlehem 23
Steinbeck, John 135
Strange Conflict 72
streams and rivers 180
Stuyvesant, Peter 181
sub-lunar realm 9
summer and winter moons 11-17
Sunset Boulevard 169-171
Susann, Jacqueline 150
Sutton, Komilla 83
Swanson, Gloria 169
Swinging Sixties 111,114

T
tango 180
Taroni, Francesca 33
Tarot 91-96
tertiary progressions 97
The Blessed Damozel 118
The Crucible 70
The Dark Side of the Moon 106
The Darling Buds of May 149
The Divine Comedy 28
The Double Helix 49
The Faerie Queene 128
The French Lieutenant's Woman 145
The Golden Dawn 88
The Hound of the Baskervilles 139
The Immaculate Conception 24
The Jungle Book 180
The Lion, the Witch and the Wardrobe 40
The Lord of the Rings 29
The Maltese Falcon 169
The Moon is Down 135
The Moon Under Water 60
The Ocean's Love to Cynthia 128

The Owl and the Pussycat 129-130
The Pentagon 182
The Picture of Dorian Gray 136
The Piper at the Gates of Dawn 107,141
The Postman Always Rings Twice 144
The White Goddess 21
The Wind in the Willows 141
The Witch of Atlas 120,121
The Witches of Eastwick 73
The Wizard of Oz 69
The Woman in the Window 169
Thessaly 66
thirteen 103,109,175,180
Thornton, Eleanor 126
Thoth 10,165
three 21-30
Three Ladies Benedight 28
Three Little Pigs 30,180
Three Marys 25
Three ships 28
Three Wise Men 23
Three witches 23
tides 5,10,44-45,122,156
Titania 127
Tolkien, J.R.R. 29
tomato 175,181
Trevelyan, Sir George 176
trilogies 29
Triple Goddess 21
tsunami 15
turkey 131
Turkey 52
Turkish flag 39

U
United Kingdom 15,60
United States 108,182
Updike, John 73
Urania 22

V
valkyries 65
Valley of the Dolls 150

Varley, John 118
Venezuela 182
Venice 182
Venus 7,169,171
Verne, Jules 161
viagra 63
Vietnam 114
Virgin Mary 24,28,173
Virginia 182
void-of-course 47

W
Waite, A.E. 93
Wallace, Alfred 38
Waltemath, Georg 161
warts 65
Washington 108,114
water 10,44,90,174,177,180,182
Waters, Roger 107
Watson, James 49
werewolf 95,180
Wesak Moon 45
Wheatley, Dennis 72
whisky 182
Wicked 70
Wicked Witch of the West 69
Wilde, Oscar 136
Wilder, Billy 169
Wilkins, Maurice 49
willow 1,141,173
Windsor forest 127
Wise Men of Gotham 25
witch 2,3,41,65-73
Witch of Endor 67
Witchcraft Act 72
wolf 30,93,180
Woodstock 111
Wyoming 183

Y
Yeats, W.B. 82,127,165
York 183

Z
Zennor 123
Zulieka Dobson 117

Also by The Wessex Astrologer - www.wessexastrologer.com

Patterns of the Past
Karmic Connections
Good Vibrations
The Soulmate Myth: A Dream Come True or Your Worst Nightmare?
The Book of Why
Judy Hall

The Essentials of Vedic Astrology
Lunar Nodes - Crisis and Redemption
Personal Panchanga and the Five Sources of Light
Komilla Sutton

Astrolocality Astrology
From Here to There
Martin Davis

The Consultation Chart
Introduction to Medical Astrology
Wanda Sellar

The Betz Placidus Table of Houses
Martha Betz

Astrology and Meditation
Greg Bogart

The Book of World Horoscopes
Nicholas Campion

Life After Grief : An Astrological Guide to Dealing with Loss
AstroGraphology: The Hidden Link between your Horoscope and your Handwriting
Darrelyn Gunzburg

The Houses: Temples of the Sky
Deborah Houlding

Through the Looking Glass
The Magic Thread
Richard Idemon

Temperament: Astrology's Forgotten Key
Dorian Geiseler Greenbaum

Nativity of the Late King Charles
John Gadbury

Declination - The Steps of the Sun
Luna - The Book of the Moon
Paul F. Newman

Tapestry of Planetary Phases:
Weaving the Threads of Purpose and Meaning in Your Life
Christina Rose

Astrology, A Place in Chaos
Star and Planet Combinations
Bernadette Brady

Astrology and the Causes of War
Jamie Macphail

Flirting with the Zodiac
Kim Farnell

The Gods of Change
Howard Sasportas

Astrological Roots: The Hellenistic Legacy
Joseph Crane

The Art of Forecasting using Solar Returns
Anthony Louis

Horary Astrology Re-Examined
Barbara Dunn

Living Lilith
M. Kelley Hunter

The Spirit of Numbers: A New Exploration of Harmonic Astrology
David Hamblin

Primary Directions
Martin Gansten

Classical Medical Astrology
Oscar Hofman

The Door Unlocked: An Astrological Insight into Initiation
Dolores Ashcroft Nowicki and
Stephanie V. Norris

Understanding Karmic Complexes
Patricia L. Walsh

Pluto Volumes 1 & 2
Jeff Green

Essays on Evolutionary Astrology
Jeff Green Edited by Deva Green

Planetary Strength
Bob Makransky

All the Sun Goes Round
Reina James

The Moment of Astrology
Geoffrey Cornelius

www.ingramcontent.com/pod-product-compliance
Lightning Source LLC
Chambersburg PA
CBHW072044160426
43197CB00014B/2619